NEO-CHINESE STYLE RESTAURANTS

新中式餐厅

深圳市艺力文化发展有限公司 编

華南理工大學出版社
SOUTH CHINA UNIVERSITY OF TECHNOLOGY PRESS
·广州·

图书在版编目（CIP）数据

新中式餐厅 / 深圳市艺力文化发展有限公司编． —广州：华南理工大学出版社，2014.3
ISBN 978-7-5623-4140-6

Ⅰ．①新…　Ⅱ．①深…　Ⅲ．①餐馆－室内装饰设计－中国　Ⅳ．①TU247.3

中国版本图书馆 CIP 数据核字（2014）第 019251 号

新中式餐厅
深圳市艺力文化发展有限公司　编

出 版 人：**韩中伟**
出版发行：华南理工大学出版社
（广州五山华南理工大学 17 号楼，邮编 510640）
http://www.scutpress.com.cn　E-mail: scutc13@scut.edu.cn
营销部电话：020-87113487 87111048（传真）
策划编辑：赖淑华
责任编辑：李　欣　孟宪忠
印 刷 者：深圳市皇泰印刷有限公司
开　　本：787mm × 1015mm　1/8　**印张**：45.25
成品尺寸：245mm × 330mm
版　　次：2014 年 3 月第 1 版　2014 年 3 月第 1 次印刷
定　　价：360.00 元

PREFACE
序言

吃是人一生中重要的生活部分，林语堂在《生活的艺术》一书中讲到中国人吃饭跟老外的区别很大。中国有句老话，叫"吃饭皇帝大"，就算是一个乞丐，他要吃饭的时候，也是比皇帝还大。只有意识到这点，设计才能明白吃饭的场所对中国人来讲是何等重要。设计师也只有了解中国人的根基以及着重点之后，方能通过设计表达出使用的原则。

中式风格的设计，与中国哲学思想互为同构关系，旨在讲究"天人合一""道法自然"等理念。中式餐厅设计是体现在民族文化上的一种哲学观念，而文化不是通过表面的摹仿和借鉴就能表现出来，它要求设计师要有深厚的文化底蕴，并且能够自然流露出来。中式餐厅设计与中式餐饮的结合，是一种血脉相承的文化。它不是一种特定的表现方式，而是一种血脉相承自然表现的形式。

我们发现张艺谋拍电影所使用的中国元素主要就三招：旗袍、大红灯笼、茉莉花。这个是声、光、色三者兼有，作为中国风电影的定式，表现得非常成熟。反观之，中式风格设计总结出来就是：虚、实、曲、直。而说到风格必然要谈到手法，手法最终是一种符号。这个符号不管是有形还是无形，最终不同的设计师表达的方式不一样。但中式风格设计在市场表现中可以概括成三种类型：纯粹的模仿，现代与传统结合，中国概念上的创新。

餐厅设计也不光是设计之后看着时尚、漂亮，更重要的是后期维护管理起来是否方便。有的餐厅设计出来，才过一两年就破败不堪，但是有的过了五六年还像新的一样，这考量的就是设计师的功力。比如说，同样的造价条件下，别的餐厅四五年就需要重新翻修，而你设计的餐厅过了八九年都不用翻新，那么这样节省下来的不仅仅是装修费用，还有人力财力物力等其他方面，这是设计的生命力，是一种价值。

什么是更高层面的附加值价值？在满足基本功能需求后，其设计独树一帜，这是其一。其二，其设计获得消费者和设计界的肯定，广受好评，连带提高店面和业主的知名度，在一定程度上强化了商业价值，这就是更高层面的附加值。国外的一些奖项设置相对比较成熟，更加具有前瞻性。对于优秀的获奖作品，他们在颁发给设计师奖项的同时，也颁给业主一个奖项，肯定业主的眼光，能够选择好的设计师，读懂好的创意设计并理解支持设计师完成这样的作品。因为一个成功的项目，离不开业主与设计师的完美配合，只有充分沟通、理解、协调，才能创作出精品。

中国地大物博，56个民族56种风格，做中式风格餐厅设计，设计师要像八爪鱼一样把触角伸向方方面面，掌握大量信息，研究中国文化。针对不同地方特色的餐厅，首先要对当地的消费市场有深入了解，其次要了解中国各地域的风俗文化、人文特点，最后是要将设计风格与当地的文化特色融合于餐厅设计中。成功的中式餐厅设计，设计创意是其中之一，该项目是否能实现最大化的商业价值是最重要的。

设计师不仅需要时时刻刻观察生活，用心思考，并要从消费者的角度，以及投资老板的角度，进行换位思考。更要赋予空间设计其艺术性，满足消费者的需求，同时提升品牌的附加值。

深圳市华空间设计顾问有限公司

Eating is an important part of human life. Lin Yu-tang talks about the great difference in eating between Chinese and foreigners in his book "Art of Living". There is an old Chinese saying: Eating is as important as the Emperor. Even a beggar, when he wants to eat, he is more important than the Emperor. Only being aware of this can designers realize how important eating places are for Chinese. And only after realizing the foundation and focus of Chinese can designers express this principle through design.

Chinese-style design, in mutually isomorphic relationship with Chinese philosophy, pays attention to theories like "Harmony between Man and Nature" and "Imitation of Nature". Chinese-style restaurant design is reflected by philosophical concept of national culture, which cannot be demonstrated through superficial imitating or learning, but requires designers to have rich cultural knowledge and can naturally reveal it. Combination of Chinese-style restaurant and Chinese food is a sort of culture with ties of blood. It is an unspecific but natural expression with ties of blood.

We can find that Zhang Yimou's films mainly use three Chinese elements: cheongsam, red lantern and jasmine. It is a combinationof sound, light and color, and is expressed skillfully as a typical form of Chinese-style films. In contrast, Chinese-style design can be concluded as: virtual, real, bent and straight. When it comes to style, skill is bound to be mentioned. Skill finally is a symbol. Whether this symbol is tangible or intangible, different designers have different ways to express it. But Chinese-style design in market performance can be summarized into three types: pure imitation, combination of modern and tradition, as well as innovation in Chinese concept.

Restaurant design is not just about fashionable and beautiful looks after design. It is more about convenient manage and maintenance in future. Some restaurants are reduced to ruin after one or two years, while some restaurants are good as new after five or six years. This is decided by designers' skills. For example, under the same condition of cost, other restaurants need renovation after four or five years, while your restaurant need no renovation after even eight or nine years, then it saves not only renovation costs, but also human and financial resources as well. This is vitality of design. It is a sort of value.

What is added value of higher level? After meeting the basic functional requirements, its design is unique. This is the first thing. For the second, its design gets approval from consumers and design field, widely-acclaimed, along with higher recognition of stores and owners, enhancing commercial value to some extent. This is higher-level added value. Some foreign awards are comparatively more mature and forward-looking. For the award-winning works, while presenting prizes to designers, they also prepare prizes for owners, affirming the owners' foresight of selecting good designers, understanding good design and backing designers' work. Since a successful project demands perfect cooperation between owner and designer, only through complete communicating, understanding and coordinating can excellent works be created.

China has vast land. Its 56 nationalities have 56 kinds of style. While designing Chinese-style restaurant, designers need to work like octopus, spreading its tentacles into every aspect, grasping large amounts of information and studying Chinese culture. For restaurants of different local characteristics, first is to have good understanding of local consumer market; then, to learn custom and culture characteristics of various regions; lastly, to integrate design style and local cultural characteristics into restaurant design. For successful Chinese-style restaurants, design creativity is one thing, while the most important thing is whether it can maximize the business value.

Designers need to not only observe life all the time with the effort of thinking, but also draw on empathy, thinking from customers and investment bosses' point of view. Moreover, designers need to give space design its artistic quality, meeting consumers' demands, and meanwhile improving brand added value.

Shenzhen Hua Space Design Consulting Co., Ltd.

CONTENTS 目录

Yeyan Restaurant

夜宴食府

Design Agency: Zhumu Decoration Design Co., Ltd. in Chongqing (L + B Chongqing Interior Design Studio) **Designer:** Li Zhou, Li Gan **Area:** 1200 m^2
Location: Yuzhong District in Chongqing **Client:** Yeyan Restaurant Co., Ltd. in Yuzhong District **Completed Date:** 2013.1 **Photography:** Chen Qinggang

设计单位：重庆筑木装饰设计有限公司（原重庆 L+B 室内设计工作室） 设计师：李舟、李干 面积：1200 m^2 地点：重庆渝中区 客户：渝中区夜宴食府餐饮有限公司
完工时间：2013 年 1 月 摄影：陈庆刚

卡斯特 品酒大师

This is a 4-story Chinese restaurant located in Yuzhong District of Chongqing. A lot of leisure entertainment places are around there. Restaurant area is not large, the layout is compact, and the modeling is concise. Metal plate and metal ceiling in the dining room are alternate with each other. LED lamps are ornamented with the elaborate golden lotus. Black and grey stones, which seem warm and stable, are used on the ground. All kinds of modern and traditional elements on the texture and color contrast constitute the internal space of the sense of rhythm, adding a new appearance to the restaurant.

这是一个位于重庆渝中区四层结构的中式餐厅，周边是重庆著名的休闲娱乐场所。餐厅面积不大，平面布置很紧凑，造型很简洁。就餐大厅深色调的仿木的金属板与金属格栅相间布置，天棚上的 LED 灯具用精心布置的金色莲花点缀，地面以黑色及灰色石材为主，空间色调整体温暖沉稳，各种现代与传统的元素形成质感及色彩上的对比，而又彼此融合，构成内部空间的韵律感，给室内增加了一种特色，给予餐厅全新的诠释。

A08

入菜口

Qiao Changnan

俏昌南

Design Agency: Donghang Decoration Interior Design Construction Co., Ltd. (Wanwei Space Design) **Designer:** Gao Bo, Zou Wei **Location:** Jingde Town
Photography: Deng Jinquan

设计单位：东航装饰室内设计建筑有限公司（万维空间设计机构） 设计师：高波、邹巍 地点：景德镇 摄影：邓金泉

The whole design is like a work of art. Elegant furniture, clean curtain cloth and special lamp give people a fresh feeling.When we are in it, we can not only enjoy the comfortable environment, but also have fun with new friends and taste the delicious dishes.

整套设计犹如一款艺术品。淡雅的家具、素净的窗帘布艺、别具一格的灯饰，给人一种清新之感。当我们身临其境，除了享受这舒适惬意的环境，也能更尽兴地和新朋好友一起品味那一道道美味佳肴。

CHINA BEAUTY

Meilin Pavilion

梅林阁

Design Agency: Xu Jianguo Architectural Interior Decoration Design Co., Ltd. **Chief Designer:** Xu Jianguo **Co-designer:** Chentao, Cheng Yingya
Area: 260 m^2 **Location:** Hefei **Photography:** Wu Hui

设计单位：中国（合肥）许建国建筑室内装饰设计有限公司 主案设计：许建国 参与设计：陈涛、程迎亚 面积：260 ㎡ 地点：合肥 摄影：吴辉

Having a quiet place to go is a happy thing. Meilin Pavilion is such a peaceful place. Time seems to have slowed down the pace here. The beautiful moments triggered everywhere. Chinese elements make the restaurant charming and unique. It not only reflects in the charm of well-proportioned ornaments, but also in the meticulously created atmosphere. Simple but delicate, ancient but novel.

拥有一个宁静的去处，是件再幸福不过的事。梅林阁就是这么一个宁静的去处。在这里，时光似乎也放慢了脚步。随处触发的是点点滴滴的美好岁月。细致的中式元素运用手法使得餐厅极具韵味。这种韵味不仅体现在错落有致的饰品上，更体现在精心营造的格局氛围中。简朴而不失精细，古韵而不失新意。

Yuemingnian Delicated Restaurant

越明年精细江南菜餐厅

Design Agency: Lonson International Commercial Design Co., Ltd. **Area:** 1500 m^2 **Location:** Futian District in Shenzhen
Materials: Marble, Gold leaf, Paint, etc.

设计单位：朗昇国际商业设计有限公司 面积：**1500 m^2** 地点：深圳市福田区 主要材料：大理石、金箔、涂料等

Everyone's mind has a picture of Jiangnan's beautiful scenery. Shenzhen Yuemingnian restaurant is the canoe coming in this southern beautiful picture, to bring a new experience of WuYue cuisine.

The restaurant serves traditional WuYue dishes. It uses modern Chinese style, in order to highlight its long tradition of catering culture and history. The use of blue and white colors makes the overall restaurant design atmosphere filled with modernism, which is pure and elegant.

每个人的心灵中都有一幅美丽的江南风景画卷。深圳越明年餐厅，便是这江南美丽画卷里驶来的一叶轻舟，给人们带来吴越美食全新的体验。

越明年餐厅提供的是传统吴越菜肴，因此本餐厅采用现代中式设计，以突出其悠久传统的餐饮文化与历史。蓝白色调的使用，使整体餐厅设计氛围充满着现代气息，格调纯净淡雅。

Xiuyifang Dining Club

秀一方餐饮会所

Design Agency: YIDUAN SHANHAI INTERIOR DESIGN **Chief Designer:** Xu Xujun **Co-designer:** Wu Yaowu **Area:** 750 m^2 **Location:** Neimenggu Naiman Banner **Completed Date:** 2013.2 **Photography:** Photography Association of Naiman Banner(Tao Zhengang, Qu Junmin, Que Ji) **Materials:** Okumatsu plate, Old wood, Rusted steel, Square pipe, Black steel mesh, Tempered glass, Stainless steel tiles, Quartz

设计单位：上海亿端室内设计 主案设计：徐旭俊 参与设计：吴耀武 面积：750 m^2 地点：内蒙古奈曼旗 完工时间：2013 年 2 月
摄影：奈曼旗摄影协会（陶振刚、曲俊民、却吉） 主要材料：奥松板、老木板、生锈钢板、方管、黑色钢丝网、钢化玻璃、不锈钢地砖、石英石

The case's new characteristic cultural theme and strong brand has attracted great attention in the local high-end catering sector. According to the location of the project, the design of overall catering space is unique and integrated perfectly in the space layout, material and lighting, color and theme, culture and management,etc, differing from general hot pot restaurant-chambers. Open the door on the first floor of the chamber like palace, what comes into sight is the desert surrounded by tamari chinensis under the blue sky and white clouds. Carefully stride cross the desert to come to the Great Wall corridor consisting of buckwheat flower, where through the gap lively dining atmosphere is dimly visible. Then, come to the dining area, and every detail of the club will inspire your curiosity, as it is so chic and elegant, fashionable and classical. Guests come here not just for having a meal to eat, more for making friends, party, business activities and so on. While gathering with friends, they can feel the regional characteristic cultural allusions. Decks on the second floor are the image of the yurts, embodying the local national culture characteristic of the project. On the third floor are high-end boxes, and each box is formed by a different theme story, interpreting the local history, culture, stories,etc.

Whole design adopts modern and simple new Chinese style. At the same time,designers borrow the local custom and culture, and refine and deduce simple materials such as homemade buckwheat flower chandelier,Great Wall corridor of Yan Kingdom with message, desert and sizzling camels at porch, staircase made of tamari chinensis full of local characteristic, palace suitcase, blue decks with the image of yurt, partition made of local plants, and distinctive Naiman prints, which all reflect the nationally fashionable and elegant temperament of clubhouse, and build a high-end food brand that is a poem. It is welcomed by the local, and consumers find the sense of belonging and have great empathy with the regional cultural inheritance.

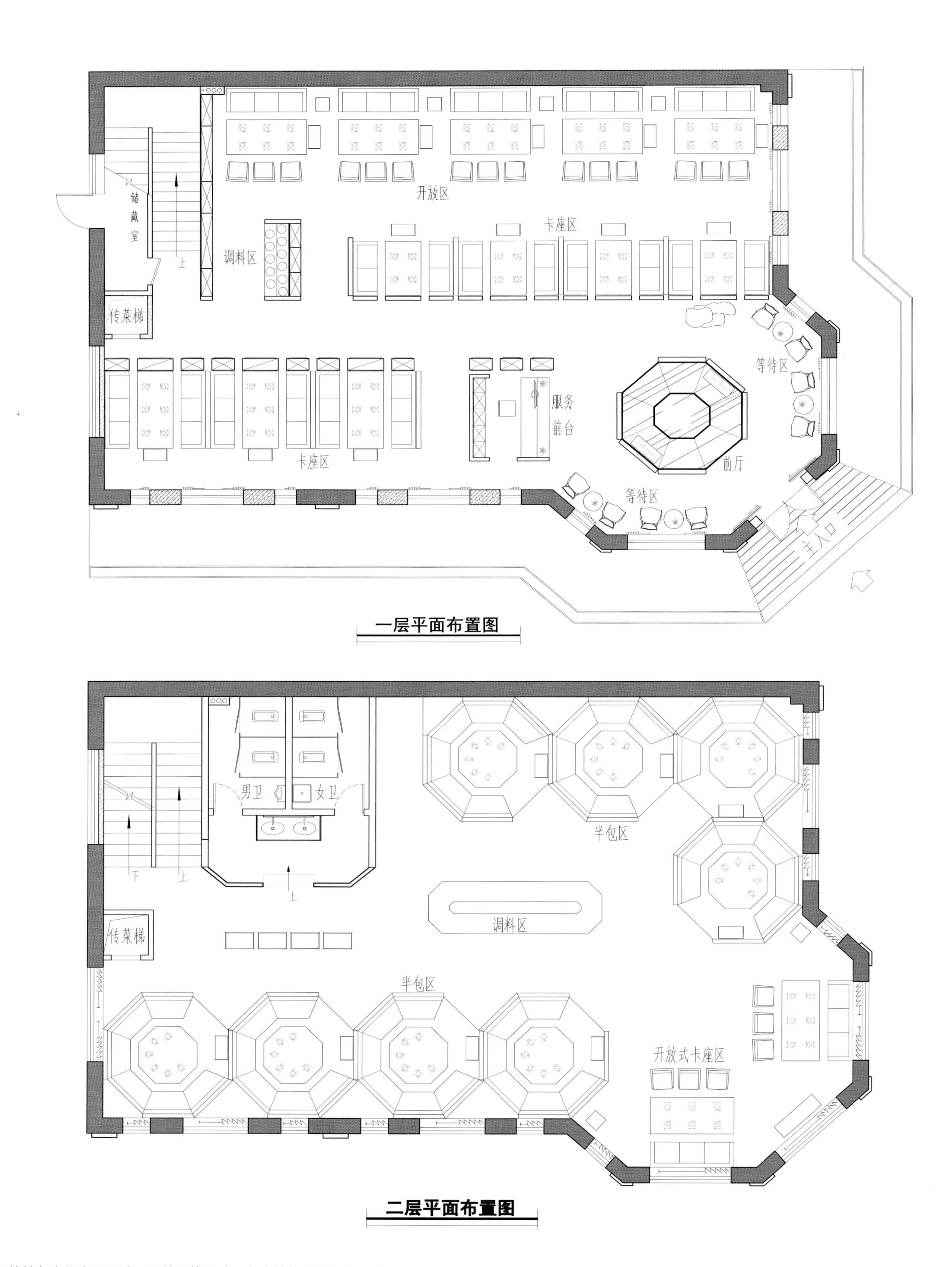

本案新颖的特色文化主题设计和品牌强势入驻，在当地的高端餐饮界引起了极大的关注。设计根据项目的定位，在空间布局、材质与灯光、色彩与主题、文化与经营等方面都独具匠心地融入到整体餐饮空间中，做到与一般火锅餐饮会所之不同，独具特色。推开王府般的一楼大门，映入眼帘的是蓝天白云下柽柳林包围的沙漠区，小心跨过沙漠来到荞麦花构成的长城走廊，透过间隙隐约可见热闹的就餐氛围，再到就餐区，会所中的每一个细节都会激起你的好奇心，别致而典雅，时尚而古典。宾客来到这里，不仅仅是为了吃饭填饱肚子，更多的是来参加交友、派对、商务等活动，聚会的同时可以亲临感受到本地区特色文化典故。二楼的半包卡座就是蒙古包的意象，体现项目的地方民族文化特色。三楼是高端包厢，每个包厢都由一个不同主题故事构成，诠释着当地的历史、文化、典故……

整个设计采用现代、简约的新中式风格，同时从当地的风土人情、地方文化挖掘中提炼演绎朴素的材料，如自制的荞麦花吊灯，具有隐寓的燕长城走廊，玄关的沙漠和铁板骆驼，用地方独特植物柽柳树打造的楼梯间，王府的提箱，蓝色的蒙古包卡座，当地植物意象隔断，以及很有特色的奈曼版画，无不折射出民族时尚高雅的会所气质，打造了一处诗情画意的高端餐饮品牌，备受当地极大的热捧，让消费者找到归属感，对地域文化传承产生思想共鸣。

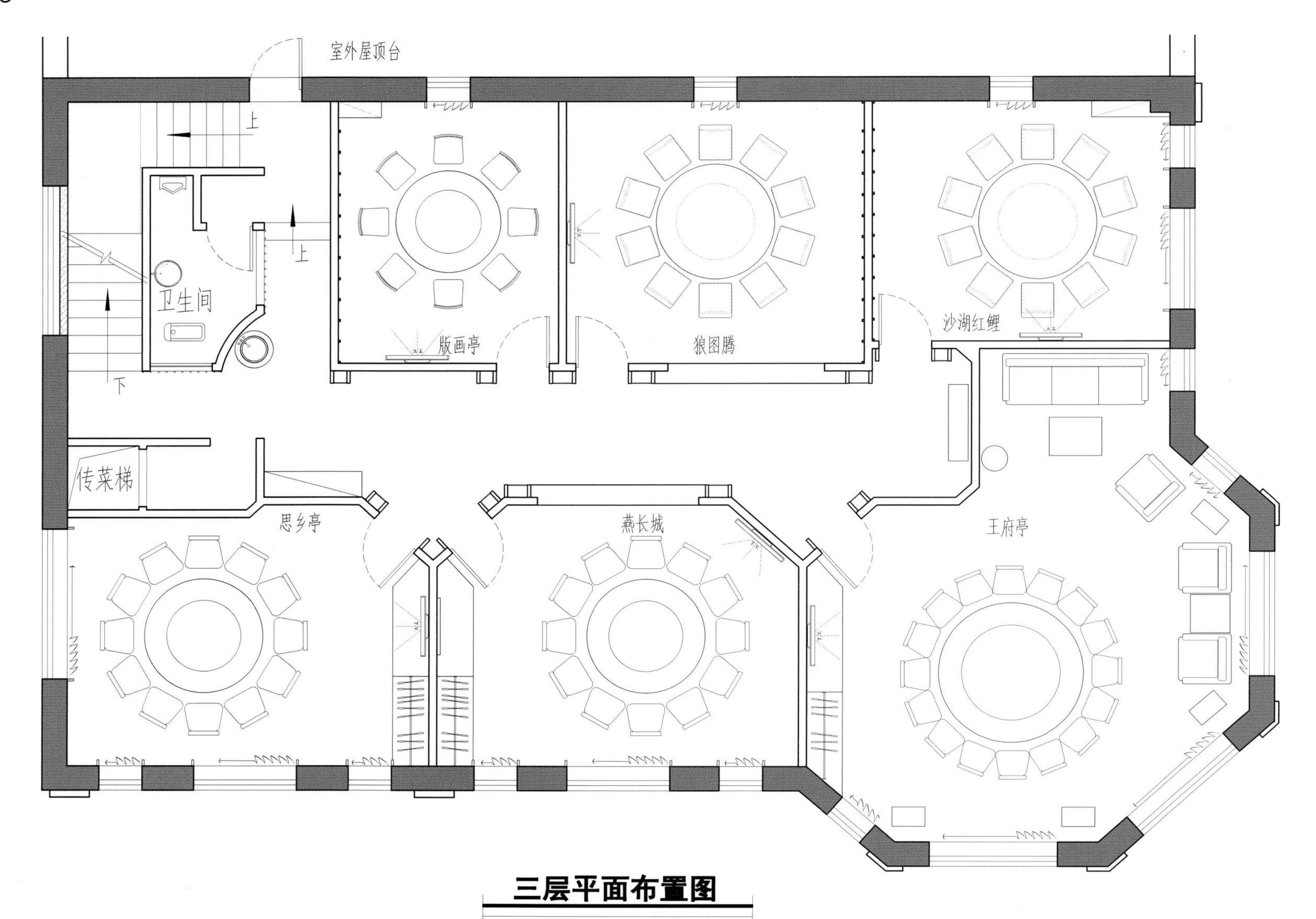

三层平面布置图

Zui Jiangnan

醉江南

Design Agency: SOKU ART DECORATION DESIGN CO., LTD. **Designer:** Chen Yonggen **Area:** 3300 m^2 **Location:** Taizhou, Zhejiang
Client: Taizhou Zuijiangnan Catering Operation Co., Ltd. **Materials:** Volcanic rock board, Cement, Preserving timber, Waterscape, Green plant, Marble, PVC pipe, etc.

设计单位：斯库装饰设计有限公司 设计师：陈永根 面积：**3300 m^2** 地点：浙江省台州 客户：台州醉江南餐饮管理有限公司 主要材料：火山岩板、水泥、防腐木、水景、绿色植物、大理石、PVC 水管等

As said by name "Intoxicated in Jiangnan", the soul of design in the project, as the main theme, is the distinction of simplicity, elegance, leisure of Jiangnan, even the rough natural volcanic rock board in the design helps show this theme.

As the main food in the restaurant is from sea and river, the elements of design theme are abstracted from there, like furred ceiling shaped as world map, lines decoration shaped into fishing net and landscaping spots of cultured stone wall, water, tree and leaf. Besides, the shared element "water" flowing in the elaborately shaped PVC pipes connects the two themes and creates the inner room of live and thought.

依据醉江南之名，该作品以江南的素雅、休闲之意境作为贯穿整体的设计生命线，粗犷、自然的火山岩板等材质的运用始终不离设计的主线。

根据此餐厅主营海鲜和河鲜的定位，从海洋和山河中提取设计主题元素，形成世界地图的吊顶，形似渔网的螺纹钢排线，文化石的墙面，水、树、叶的造景，并运用两主题"水"的共性，利用水管的变异造型将其连接、互通，形成有生命、有思想的室内空间。

火警电话 119

309

Flower and Fish in Xinjiang

新疆花枝沸腾鱼

Design Agency: Xu Pin Design and Decoration Engineering Co., Ltd. in Suzhou **Chief Designer:** Jiang Guoxing (Jonny) **Co-designer:** Tang Zhennan, Li Haiyang, Han Xiaowei **Area:** 850 m^2 **Location:** Urumqi, Xinjiang **Completed Date:** 2012 **Materials:** Wood veneer, Wood, Stone, Litchi surface treatment, Grey mirror, Wood flooring, Seaweed mud

设计单位：苏州叙品设计装饰工程有限公司　主案设计：蒋国兴　参与设计：唐振南、李海洋、韩小伟　面积：850 m^2　地点：新疆乌鲁木齐　完工时间：2012 年
主要材料：木饰面、木格、石材、荔枝面处理、灰镜、木地板、海藻泥

This project is located in the ancient Silk Road, new North Road in Urumqi City, the modern fashion sense and Oriental elements embedded in the whole space, contains profound artistic conception of Oriental atmosphere. Designers draw inspiration from the traditional Oriental elements, exquisite blue and white porcelain pottery basin, and making here charming. Designers use color to perfection in one's studies. Thick accessories in drawing paintings makes space ventilation, vitality. The traditional Oriental culture is not a simple list, but by contemporary design form, language, thinking to express the contemporary aesthetic temperament. In this full of imagination space, contemporary art and traditional culture encounters, art and space collision, life is full of details in the space, and has a romantic temperament, becoming your mental conversion zone.

本案位于古丝绸之路新北道重镇乌鲁木齐，简约的现代时尚感与东方元素深植于整个空间之中，蕴含大气深邃的东方意境。设计师从传统东方元素中汲取灵感，精致典雅的青花瓷陶盆，意趣盎然。设计师炉火纯青地运用厚重色彩，加以配饰的白描挂画使得空间透气，生机勃勃。这些传统东方文化绝不只是简单罗列，而是通过当代设计形式、语言，张扬地表达当下的审美气质。在这个充满想象的空间里，当代艺术和传统文化邂逅，艺术与空间碰撞，生命在空间充满盈动，而拥有一种浪漫主义的气质，成为你心理皈依的地带。

灭火器箱
火119警

Man

Chenshijiu Pot — Novel Chinese Food

陈仕玖煲——创意中国菜

Design Agency: Senpeng Interior Design Co., Ltd. in Xi'an **Designer:** Shi Yansen **Area:** 1200 m² **Location:** Xi'an **Completed Date:** 2012. 9
Materials: Italian pottery rustic tiles, NVC, Finished teak veneer, Wood stone

设计单位：西安森鹏室内设计有限公司 设计师：石炎森 面积：1200 m² 地点：西安 完工时间：2012 年 9 月 主要材料：金意陶泛古砖、雷士、成品柚木饰面板、木纹石材

The works is with the idea of the original. The irregular steel frame at the top is a form of decoration, while effectively highlights the overall personality. Ground uses deep tones antique bricks. It was decorated with materials such as steel, iron ornaments, paintings and other accessories. The idea of the top floors, walls and accessories are well in line with the design theme.

作品构思以原创为主。顶部的不规则钢架既是一种构成装饰，同时又有效地突出整体的个性。地面采用深色调仿古砖，统一整体。在配饰上利用钢筋、铁屑等一些常见材料做挂饰、挂画等。顶面、地面、墙面以及配饰的构思都很好地切合了设计的主题。

陳仕玖煲
創意中國菜

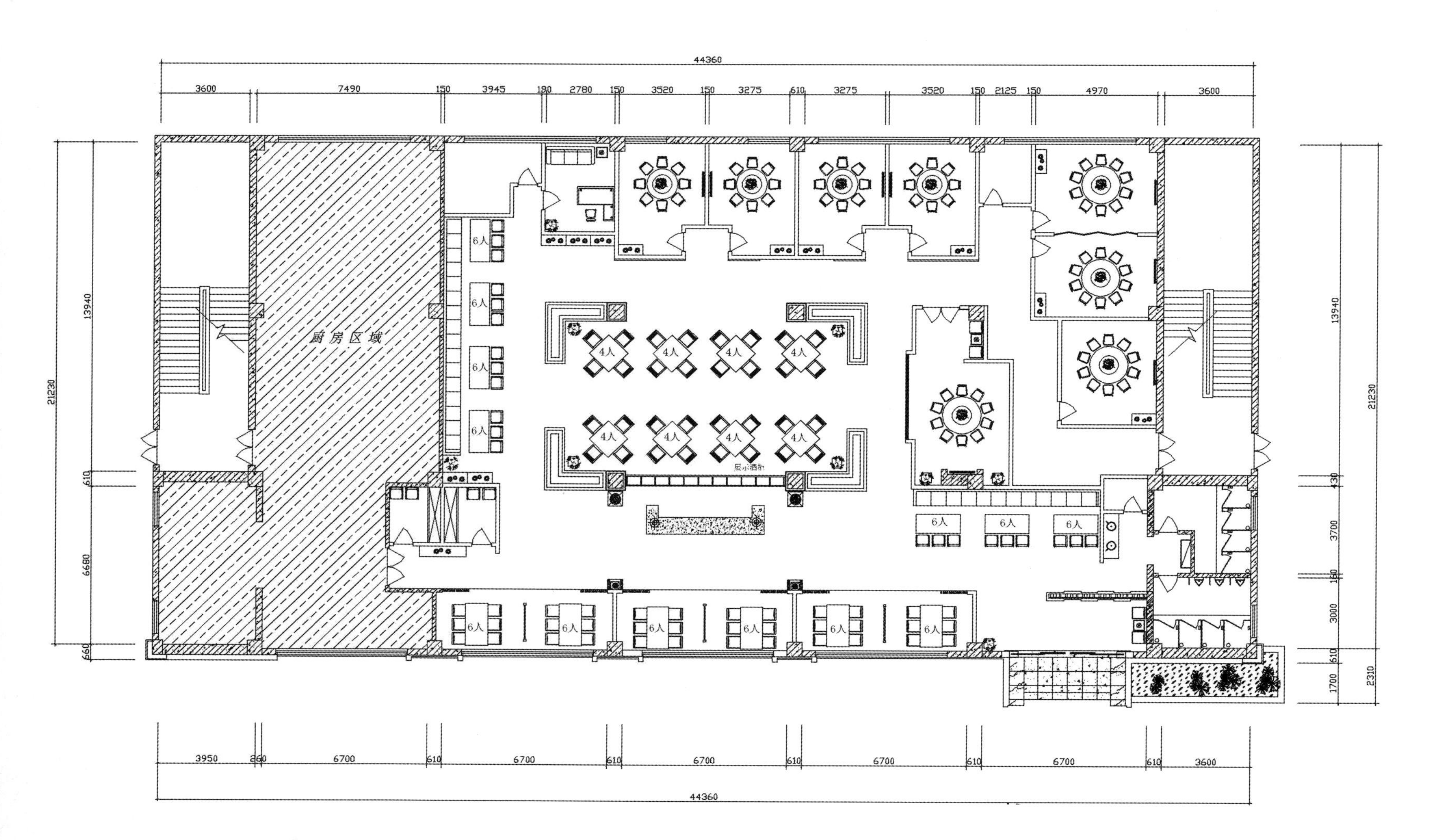
44360
3600
7490
150
3945
180
2780
150
3520
150
3275
610
3275
3520
150
2125
150
4970
3600
厨房区域
6人
4人
13940
21230
610
6680
660
430
3700
160
3000
610
1700
2310
3950
860
6700
610
6700
610
6700
610
6700
610
6700
610
3600
44360

The Pot, Zhongshan Road

这一锅 中山路店

Design Agency: JOY INTERIOR DESIGN STUDIO **Chief Designer:** JOY-CHOU YI **Co-designer:** Chen Zhiqiang **Area:** 380.7 m² **Location:** Taipei, Taiwan **Completed Date:** 2012.11 **Photography:** Lv Guoqi, Gentle Breeze Photography **Materials:** Culture stone, Marble, White cement + straw, Sandblasting glass, Backoak, White steel brush sycamore

设计单位：周易室内设计工作室　主案设计：周易　参与设计：陈志强　面积：380.7 m²　地点：台湾台北市　完工时间：2012 年 11 月　摄影：和风摄影吕国企
主要材料：文化石、大理石、白水泥 + 稻草、喷砂玻璃、黑橡木洗白、钢刷梧桐木

"The Pot"is located in Zhongshan Road of Taipei City, managed by 85℃ .

The restaurant costs forty million yuan, attracted a lot of attention, the driving force is JOY-CHOU YI who has years of business experience. JOY-CHOU YI uses China Eastern representative elements, such as carved, window grilles, motif and cucurbit, to create a palace-like dining space.

The entrance uses Chinese style cross grille to attract the passengers' curiosity. The appearance of the store is eye-catching and the interior space is also impressive. Oriental antiques reception with culture stone wall to dilute the sense of gorgeous with natural elements, so that to release the air of relaxation.

No matter in the design and practical can let visitors experience a journey such as the Royal.

位于台北市中山北路的“这一锅”，由知名企业 85℃ 跨足经营。

耗资四千多万打造的用餐环境，引来不少人注目，其背后推手正是具有多年操作商业空间经验的周易。周易采用雕花、窗花、花片、葫芦等这些具有中国东方代表的元素，营造一处具有如宫廷意境的餐饮空间。

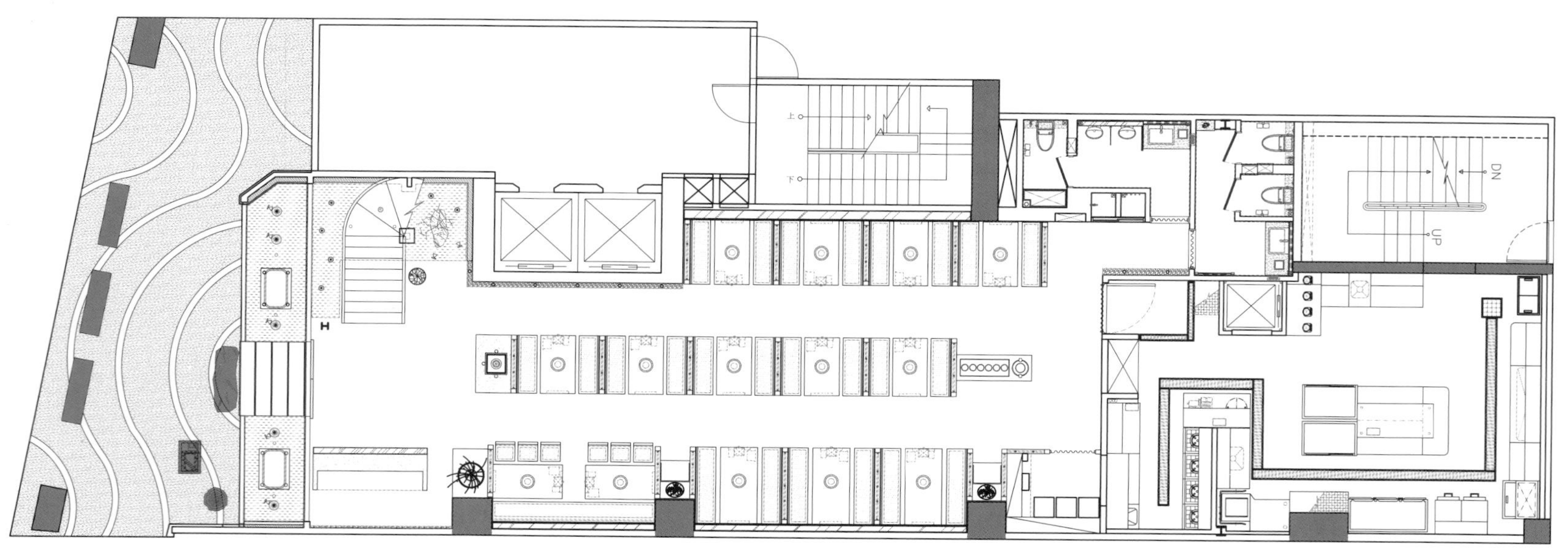

入门处，则采用中国风的十字格栅，为的就是让路过的行人产生某种亢奋与好奇，进而诱发入内消费的冲动。店面外观不仅引人注目，内部空间同样令人惊艳。东方古董般的接待柜台搭配文化石背墙，试图用自然元素来冲淡华丽感，让空间释放出一点休闲放松的气息。

不管在设计上还是实用上，都能让来客体验到一场如皇室般尊贵的感官之旅。

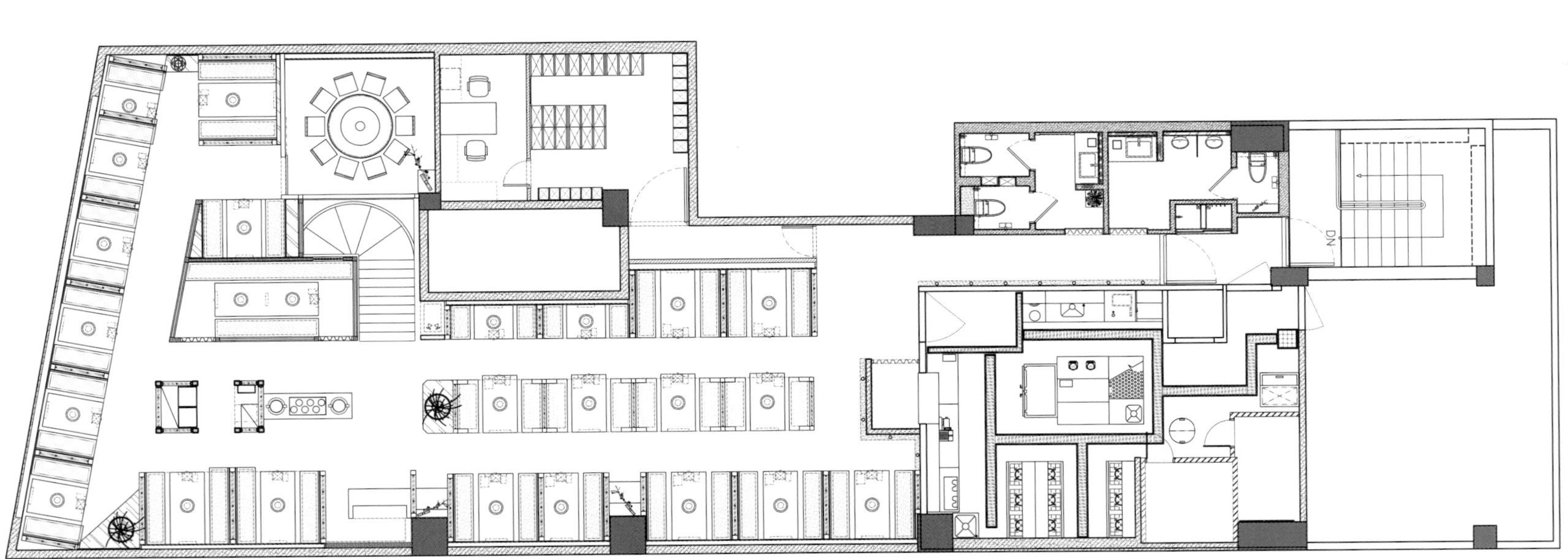

Moonlit Lotus Restaurant

荷塘月色餐厅

Design Agency: Yijing Interior Design Agency in Chengdu **Designer:** Zhong Qiming, Huang Ying, Xu Zhou **Area:** 4300 m^2
Location: Suining, Sichuan **Client:** Yang Chun Ba Ren Restaurant **Completed Date:** 2012.9 **Materials:** China black stone, USA hickory, Tinted hlass brick, Imitation marble lamp box plate, Stainless steel mirror, Brushed stainless steel, Art glass, Antique floor tiles, Installation art lotus, Batten

设计单位：成都易景室内设计事务所 设计师：钟其明 、黄莺、徐舟 面积：**4300 m^2** 地点：四川遂宁 客户：阳春巴人食府 完工时间：**2012** 年 **9** 月
主要材料：中国黑石材、美国胡桃木、茶色玻璃砖、仿云石灯箱片、镜面不锈钢、拉丝不锈钢、艺术玻璃、仿古地砖、装置艺术莲花、木条

This project achieved the overall sense of quality from the pre-planning to restaurant decoration. The owner hopes to build a new type restaurant with culture and art, which not only has the space of the fashion style, but also has profound Confucianism elegant cultural atmosphere.

The restaurant used rare Song porcelain lotus leaf cover pot and Buddhism culture as the main line. The overall style is concise and easy, and the color is dense but elegant. Outdoor and indoor are well- unified. Light is changeable, the pattern is tailored to each space, integrated into the design theme elements, added the dining atmosphere, created a high-quality cultural appeal.

本案通过从前期策划到餐厅装饰设计的整合来达到餐厅整体的品质感，业主希望打造一个全新概念的文化与艺术型餐厅。既有时尚的空间风格，又有深厚的儒雅人文底蕴。

餐厅以稀世珍宝宋瓷荷叶盖罐和佛教中的观音文化为设计主线。整体风格简洁、大方，色彩浓重不失高雅；室外及室内融为一体，用材统一、整体。灯光富于变化，样式也似为每个空间量身定做，融入了设计主题元素，增添了室内用餐氛围，营造了高品质的人文情调。

Kaixuanmen 7th Club

凯旋门七号会馆

Design Agency: DING HE DESIGN in Henan **Chief Designer:** Sun Huafeng, Liu Shiyao, Kong Zhongxun **Co-designer:** Li Chuncai, Wang Fenli, Hu Jie, Sun Jian, Shi Yunxiang, Yang Jingrui **Area:** 5000 m^2 **Location:** Luoyang **Materials:** Hydrological sandstone, Italy wood-grain stone, Portopo, Black titanium

设计单位：河南鼎合建筑装饰设计工程有限公司 主案设计：孙华锋、刘世尧、孔仲迅 参与设计：李春才、王粉利、胡杰、孙健、师云香、杨景瑞 面积：5000 m^2
地点：洛阳 主要材料：水文砂岩、意大利木纹石、黑金花石材、黑钛金

Luoyang is an ancient capital city of nine dynasties. In such a city of rich history and culture, how to combine business operations, customer experience, current design trend with the rich regional culture to bring the customers remarkable noble experience is the focus of the consideration at the initial stage of positioning.

Entrance is the first port customers experience, and stone segmentation of the clubhouse's facade in graceful disorder well handles a feeling of grandeur created by the big building volume, which brings strong visual impact and at the same time saves the cost. The combination of red and black grille and the door jamb highlights the entrance position and also forms the visual sense of receiving guests.

Entering into the hall of the clubhouse, stone pillars, tall red wall lamps, lattice of Chinese style, and black gold-outline lacquer cabinets, form strong ritual sense and highlight the dignity of the guests. What shock people more are two high atriums of eight meters: brilliant petals above the atrium dead against the entry pour down and converge into a glittering and translucent a crystal peony, imperial but gentle, which shows the theme of "Louyang is the home of peony". The small atrium on the left of the hall acts as the rest area as well as treasure cabinet. There, happy color figurines of high imitation naughtily stand in two sides; concise arhat bed under the whatnot soaring to the top is for a short rest. The entire space is surrounded by rich culture atmosphere but also penetrating and lightsome.

On the first floor, besides characteristic compartments of different styles, there is the area of 240m^2 for extra seats, prepared for sporadic guests or business activity to meet the needs of various customers and at the same time make the space more smart and humanized. From the first floor to the

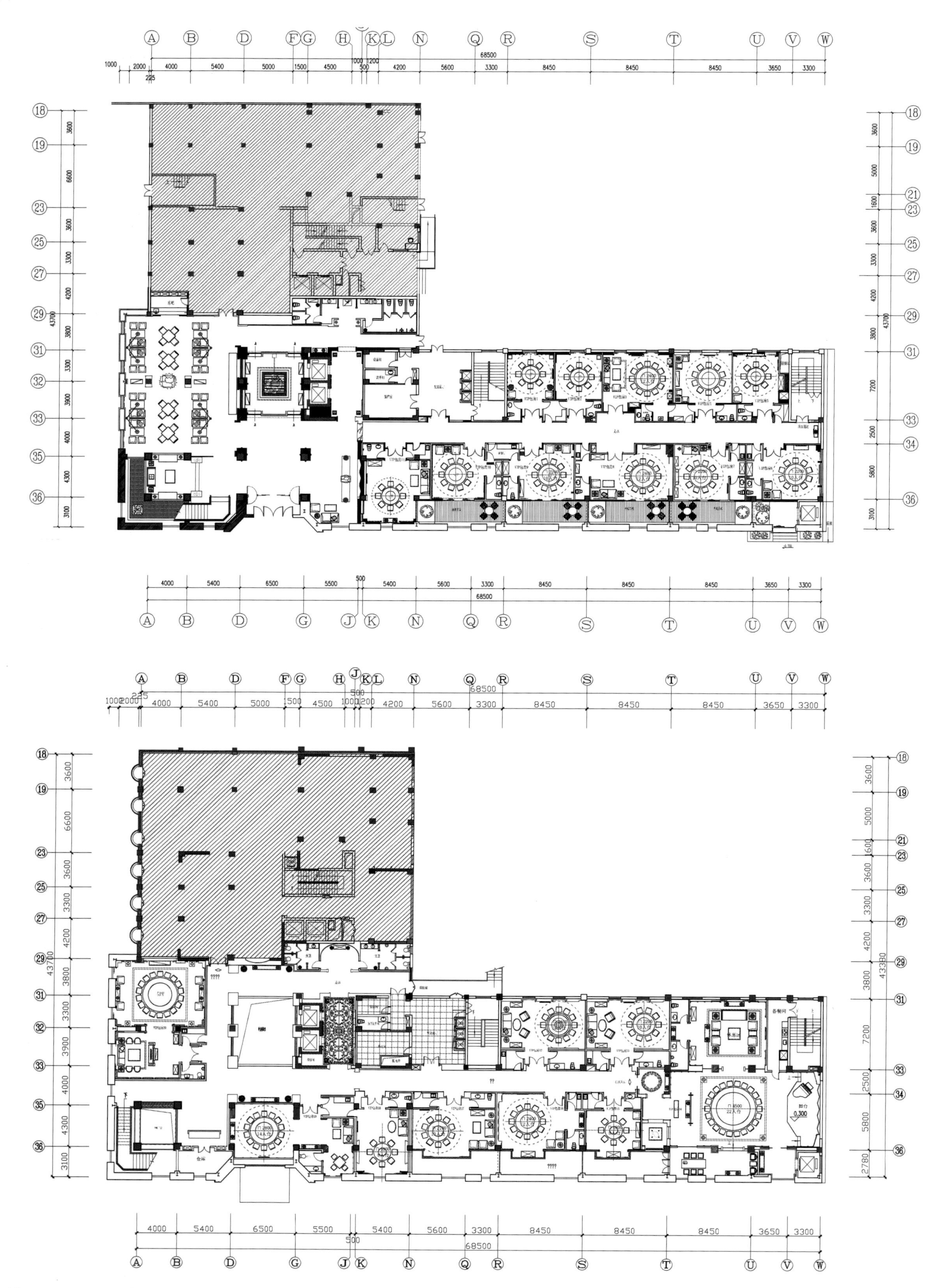

second floor you can use elevator or walk by the step ladders around the treasure cabinet, the ground paved by hole stone leads people into restaurant rooms on the second floor. There are some featured rooms on the second floor , of which east palace(luxurious rooms of Chinese style), west palace (woman's club of neo-classical style), Islamic room are most unique, to show mightiness and elegant atmosphere of the space and meet personality needs of distinguished guests, high-end female guests and Islamic guests , embodying the aim of the humanized service.

From layout arrangement to adornment skill and detail description, designers give top priority, and they make rich regional culture naturally flow in space, abandon the complicated Chinese symbol, reserve inclusiveness and mightiness of the culture of the central plains and combine aesthetic interest of modern people to bring people a feast of collisions of different cultures.

在九朝古都洛阳这样一个历史文化积淀深厚的城市，怎样将商业运营、客户体验、当下的设计潮流与厚重的地域文化巧妙地融合，给客户带来非凡的尊贵体验是在初期定位时考虑的重点。

入口是客户体验的第一站，会馆外立面错落有致的石材分割很好地处理了建筑体量过大带来的沉重感，带来强烈视觉冲击的同时也很好地节约了造价。红黑相间的格栅和

门套的结合强调了入口位置，也形成吸纳迎人的视觉感受。

进入会馆大厅，通道两侧的石材柱子、高大的红色壁灯、通顶的中式花格、黑色描金漆柜等元素形成了强烈的仪式感，凸显客人的尊贵，更让人震撼的是两个挑高8米的中庭：正对入口的中庭上方璀璨的花瓣倾泻而下汇聚成晶莹剔透的水晶牡丹，气宇轩昂却也柔情似水，体现出洛阳作为牡丹之都的主题。大厅左侧的小中庭则作为藏宝阁休息区，高

仿的唐代彩乐俑顽皮地立于两侧，高耸至顶的古董架下简洁的罗汉床可供人短暂休憩，整个空间都被浓浓的文化气息所包围却不乏灵动通透。

一层除了风格各异的特色包间外，还为零星客人或企业活动准备了一个 240 m^2 的散座区，满足各类客群需求的同时也让空间更加灵动和人性化。人们可以从一楼搭乘电梯或从藏宝阁的步梯步行到二楼，洞石铺就的地面将人们引入二层餐厅包间。二层的包间中有几个特色包间，其中东宫（中式风格豪华包间）、西宫（新古典风格女性会所）和伊斯兰包间最具特色，呈现出尊贵大气、细腻典雅的空间氛围，满足了贵宾接待、高端女性客户及伊斯兰民族客人的个性需求，体现其人性化服务的宗旨。

从平面布局到装饰手法再到细节刻画，设计师举重若轻，使得厚重地域文化在空间中自然流淌，摒弃繁复的中式符号，保留中原文化的包容大气，融和现代人的审美情趣，带给人一场不同文化碰撞的尊贵飨宴。

Happy Village Restaurant

聚乐村饭庄

Chief Designer: Li Ming **Co-designer:** Yao Cui, Wang Zhenhua, Zhang Shuai **Area:** 1600 m^2 **Completed Date:** 2012 **Photography:** Li Ming

主案设计：李明 参与设计：姚翠、王振华、张帅 面积：**1600 m^2** 完工时间：**2012** 年 摄影：李明

Happy Village Restaurant is located at the Boshan. It is the catering intangible cultural heritage item.

Scheme adopts traditional Chinese courtyard layout form. It combines the traditional elements and modern technology, in the inheritance and development. Found a kind of integrity expression peculiar to the northern culture atmosphere, carefree, open, honest, plain, constructs the environment space's characteristics. Space is rhyme but not gorgeous, bright but not breakable.

Portal type steel structure of the echo stretching everyone sculpture to the courtyard to the local space of indoor decoration, which show the theme culture.

To a restaurant, with Confucianism, Buddhism, Taoism culture as the breakthrough point, it will integrate shadow shapes into shape element, make the old brand happy village into a museum that can provide people with food.

聚乐村饭庄位于鲁中博山，为餐饮中的非物质文化遗产项目。

方案采用中国传统的四合院式的布局形式，设计形式将传统元素与现代工艺结合，在传承中又有发展。寻着一种气节表达北方特有的文化氛围，大气、畅快、舒展、浑厚、质朴，构造了环境空间的特质。空间韵而不艳丽，空间明而不失志。

入门的钢构造型福字雕塑到院落的呼应延伸到室内局部空间的饰品，都展示出了主题文化讲究的一面。

以餐饮为龙头，以儒释道文化为切入点，将光影造型融入造型元素中，力求把聚乐村老字号打造成能吃的博物馆。

一九一九年
聚樂村

始于一九一九年

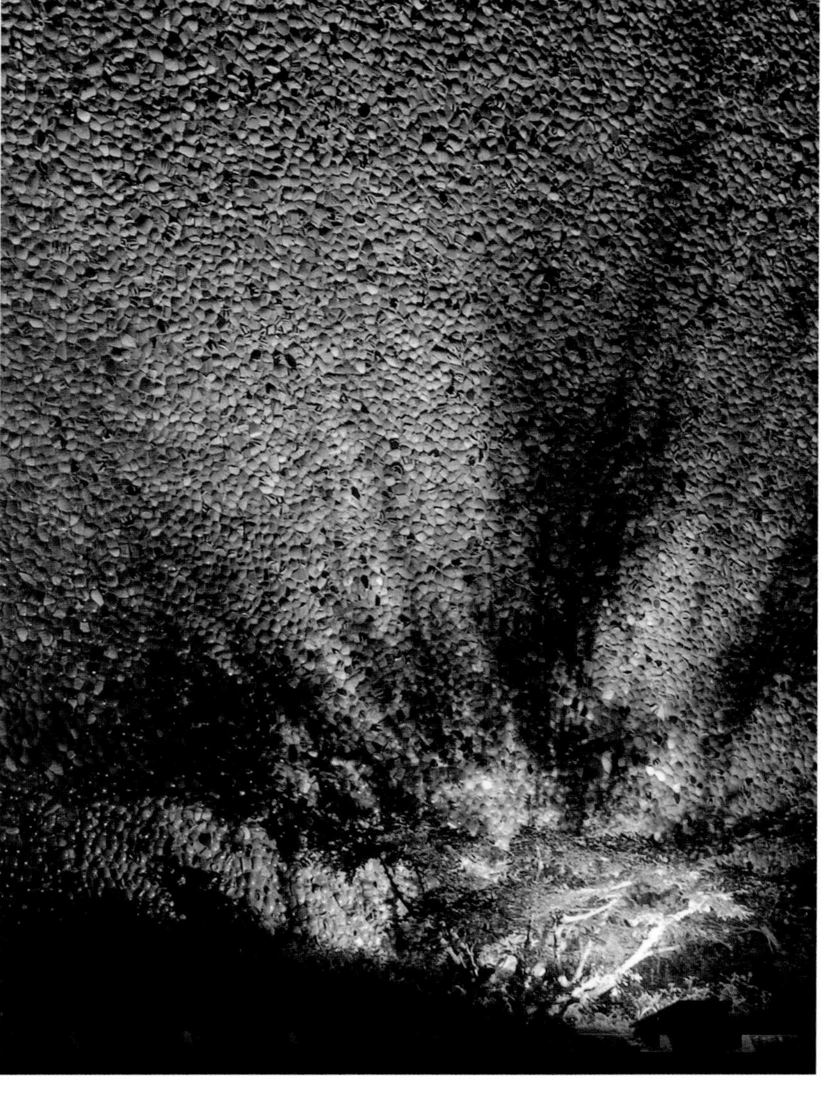

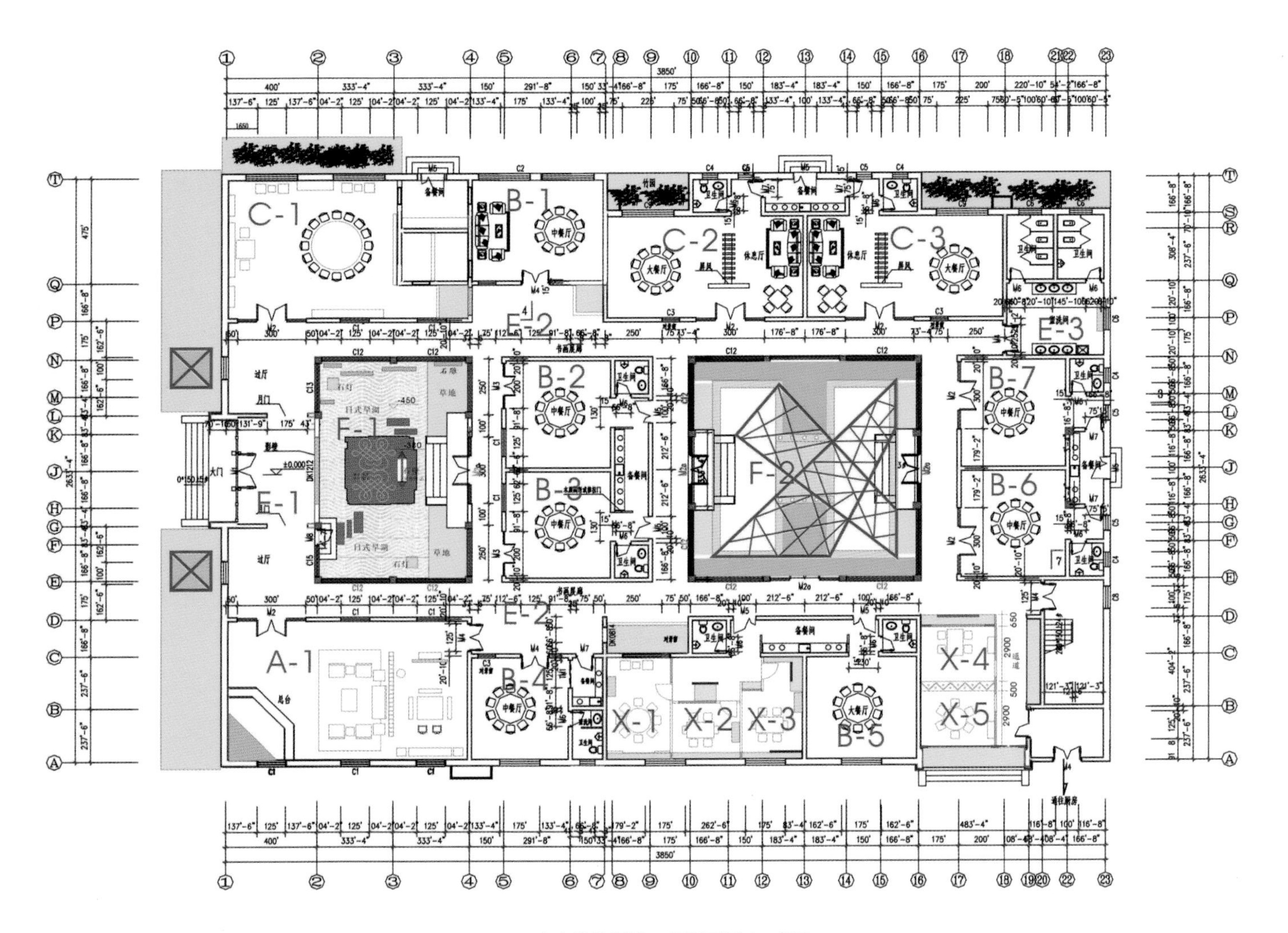

室内装饰设计一层平面图 1：200

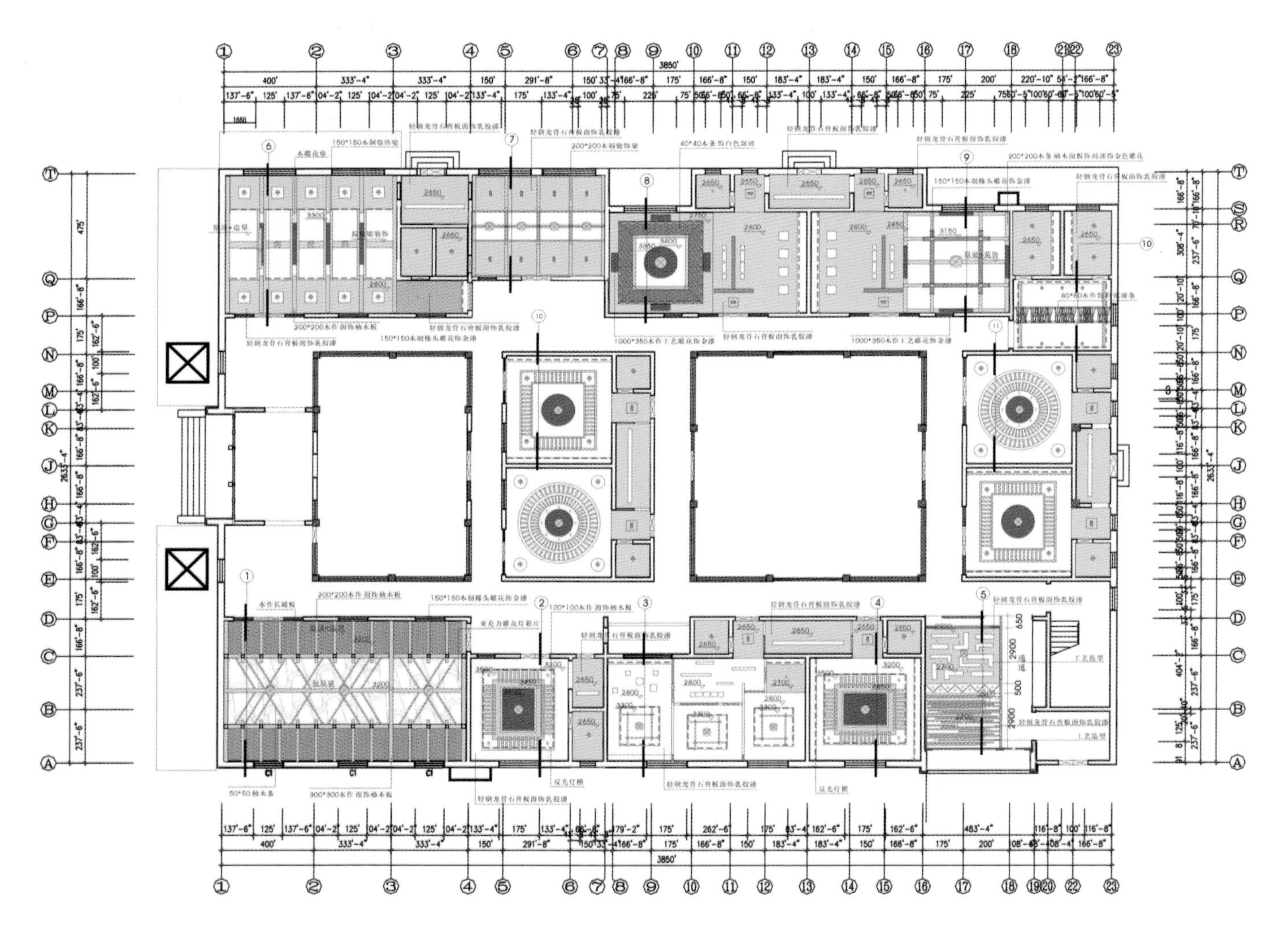

室内装饰设计一层吊顶平面图 1：200

聚樂村

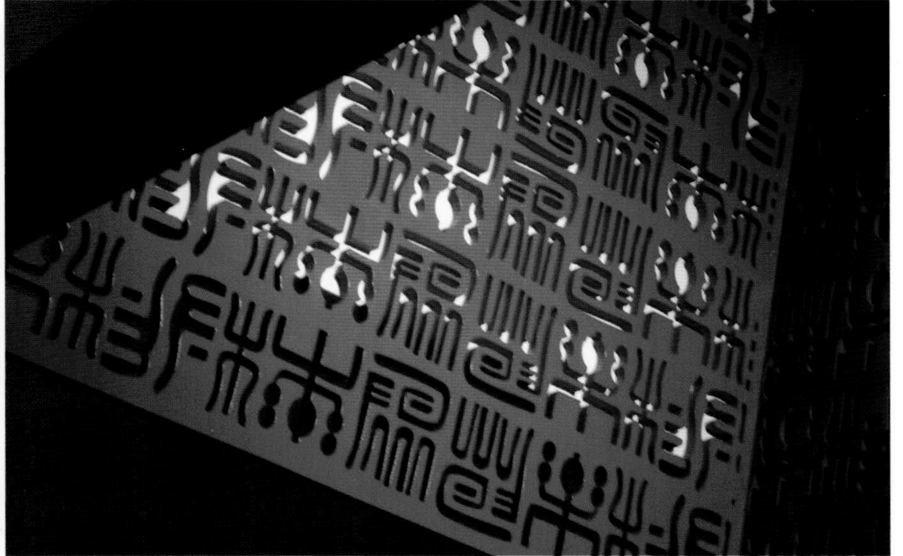

福
祥

Private Club Restaurant in Beijing

北京近郊私人会所餐厅

Design Agency: Atelier de L'A.V.I.S in Beijing **Chief Designer:** Cai Zongzhi **Co-designer:** Zhong Zhiding, Tong Xiaoyou **Location:** Outskirts of Beijing **Completed Date:** 2012.12 **Photography:** Sun Xiangyu **Materials:** Old wood, Stone, Glass, Square tube

设计单位：北京法惟思设计 主案设计：蔡宗志 参与设计：钟智鼎、童孝友 地点：北京近郊 完工时间：2012 年 12 月 摄影：孙翔宇
主要材料：旧木料、石材、玻璃、方管

In this case, designers put the surrounding landscapes into it which makes the interior space more sunlight, green and healthy. Meanwhile, the number of elements (walls, parapet walls, hedges, ponds, etc.) has been increased to make the indoor and outdoor space rich, emphasizing space and the environment are inseparable.

Therefore, during the construction phase, workers need watch all angles of the outer, adjust the level of the actual situation of the spatial elements, weaken the landscape inside out inappropriate, and strengthen the integration of indoor and outdoor space.

There are two paths to entering the club. The inside entrance has set up a guest room. The outside entrance is a corner of the landscape, which is near the parking lot. The poncho extension of the concept of the screen wall blocks the parking lot, leaving the beautiful sky and branches.

The display of study room represents the master's mind and culture, so as to communicate with guests. Elements of traditional Chinese roof are not a show of complex wooden structure, but the beauty of reversal. Study space surrounded by a water circle. It has ornamental koi and boulders.

The dining room at the end of the sequence is the most important space to entertain guests. Large lamps become interesting space furniture. The bookshelf screen is made of black steel material.Teppanyaki room intimates exchange between the host and the guests. The bookshelf light is the focus of the space.

Backyard is the most difficult place to deal with. It is a space for a variety of things. It too closes from the dining room, so a few simple wall and pool make it different immediately. If we put a few big rocks and big vat in it will be better.

该案例在空间设计处理上，注入了周围的景观让室内空间更阳光、绿色和健康，同时增加了一些元素（高墙、矮墙、绿篱、水池，等等）让室内外空间丰富了起来，强调了空间与环境的不可分割。

因此这个项目在施工阶段时，设计师需不断地观察空间外部的各个角度，调整空间元素的高低与虚实，弱化外面不合适的景观入内，强化室内外空间的融合。

进入会所有两个路径，内入口需经过一个迎宾空间，外入口是内院散步景观的一角，离停车场较近，入口的雨披设计是影壁墙概念的延伸，与户外的景观墙共同挡住停车场，留下美丽的天空与树枝。

书房的陈列代表的是主人的心思与文化，借此与宾客交流。传统中式大屋顶的元素，在这不是展示木结构的复杂之美，而是反转处理外屋顶轮廓线的美，书房空间的周围是一圈服务动线结合水系，水系内有观赏锦鲤及大石。

用餐空间是空间序列的末端，也是主人招待宾客的最重要空间。大型的灯具成为空间中有趣的一个空间家具。旁边的书架屏风使用的是黑钢的材质。铁板烧包间体现了宾主之间的亲密交流，书架灯是空间的重点。

后院，一般是最难处理的地方，它是各种不知如何搁置的东西的空间，这里离用餐地方太近了，所以简单的几个墙、框架及水池，立马会有不同的视觉效果，如果再摆几个大石块及大瓮，那就更好了。

福
茶

Number One in Su Garden

苏园一号

Design Agency: DING HE DESIGN in He'nan **Chief Designer:** Liu Shiyao, Sun Huafeng **Co-designer:** Li Xirui, Sun Jian **Area:** 1510 m^2
Location: Zhengdong New District **Materials:** White wood stone, Ancient wood stone, African rosewood, Mat series, Hard package, Bricks, etc.

设计单位：河南鼎合建筑装饰设计工程有限公司 主案设计：刘世尧、孙华锋 参与设计：李西瑞、孙健 面积：1510 m^2 地点：郑东新区
主要材料：白木纹石材、古木纹石材、非洲紫檀、席编、硬包、城砖等

The original building is a two-story modern sales center. The designer consciously sets the entrance behind the courtyard and the corridor, making the visitors have a feeling of delicateness and connotation of Chinese garden. The reconstructed building is applied to Hui architectural form. The room is designed to reflect the literal inspiration of Chinese gentlemen.

In the display design, it mixed the new western classical and new Chinese style, not only has a good comfort, but also achieves excellent visual effects. Using blue linen makes the warm color of large wood rattan to be well balanced, curtains let guests have a sense of enclosure and break the crudeness of the traditional timber buildings.

原建筑为两层结构现代风格的售楼中心，设计师有意识将会所入口设计为通过庭院进入廊道再进入会所，让客人有节奏的感受到中国园林的含蓄和精致。改造成的建筑为简约的徽派建筑形式。房间的设计以书架为依托体现出中国文人雅士的情怀。

在陈设设计中，采用西方新古典及新中式家具的混搭，既有了很好的舒适感又取得了极佳的视觉效果。宝蓝色布草的运用使大面积木色藤编的暖色得以很好的平衡，布幔和竹帘让在散台区的客人既有了围合感又打破了传统木构建筑的生硬。

蘇園

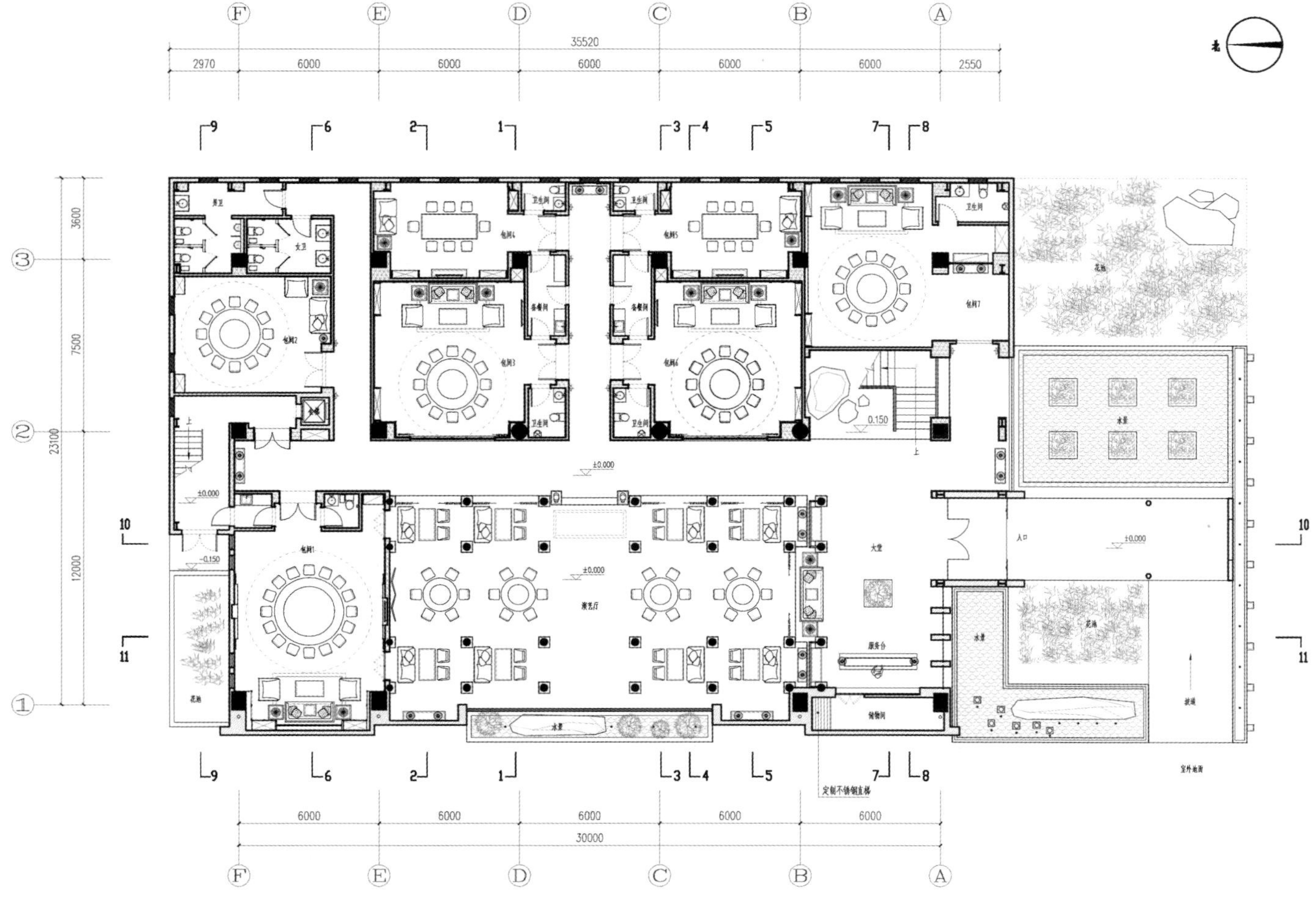

一层总平面布置图 1:100

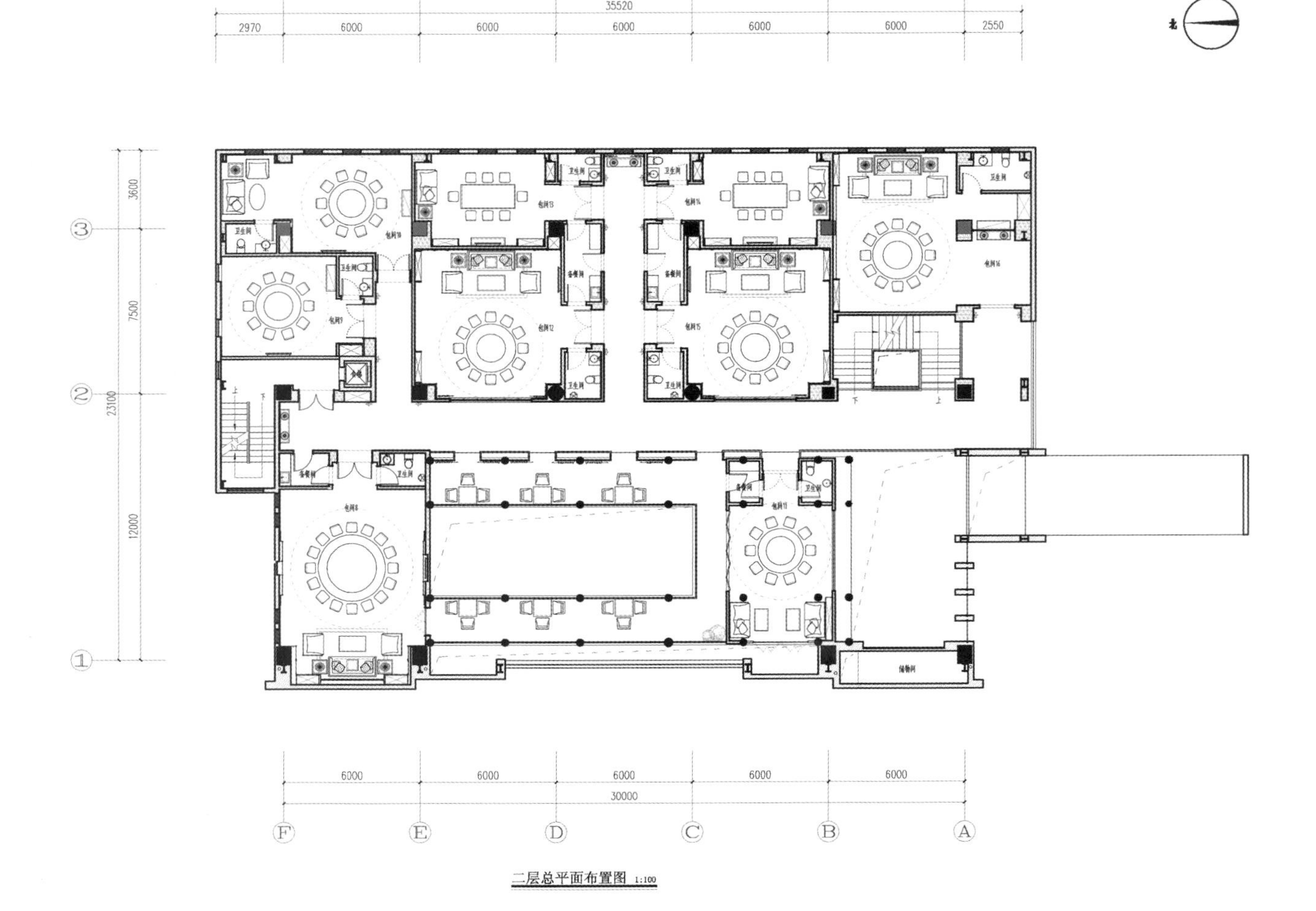

二层总平面布置图 1:100

A Feast at Yuanshanhui

圆山会一席宴

Design Agency: The Gazer Design Center in Hangzhou **Designer:** Lin Sen

设计单位：杭州肯思装饰设计事务所 设计师：林森

The restaurant was designed by architect Lin Sen. A Feast as a restaurant focuses on food culture of Nanxun, and combines regional technology and art. Nanxun mixes the southern delicateness and the northern force without conflict. It is very difficult to grasp the balance between them.

餐厅设计由设计师林森主笔，因为一席宴本身就是餐馆，设计的切入点就是做南浔的美食文化，把地域性的工艺和艺术结合进去。南浔既有南方的精巧、细致，又有北方的大气、雄浑，两者融合，毫不冲突。设计要做到既大气又精细，是很难把握的。

River Restaurant

井河公馆

Design Agency: Dashidai Design & Consulting Co., Ltd. **Designer:** Wu Xiaowen, Zhang Yingjun **Area:** 2000 m^2 **Location:** Konggang, Tianjin
Completed Date: 2011.1 **Photography:** Xing Zhentao **Materials:** Wood floors, Ebony, Wood, Air brushing

设计单位：大石代设计咨询有限公司　设计师：吴晓温、张迎军　面积：2000 m^2　地点：天津市空港　完工时间：2011 年 1 月　摄影：邢振涛
主要材料：木地板黑檀、木料、喷绘

Salt merchants culture is the backbone of the design, by which the design thought is spread . Most people in Yangzhou live in trading salt, and they're accustomed to flashiness and skillful in delicacies, so Yangzhou's feast is famous around the country. The food specification formed by the salt traders and salt officers is the style of massive spectacle,elegant environment, peculiar dish, perfect material selection, exquisite feeder.

First, the essence of Huaiyang cuisine. Taking taste as the core and health as purpose, is extended to the combination of virtual reality and real scene "taking landscape as core and pleasantness as purpose"(wall paintings of picturesque Yangtze and small bridge over the flowing stream complement each other and form a beautiful scene). Second, the salt merchants' rich and elegant delight of life. The "static", "elegant", "interest" of life space, through the arrangement of landscape, make the guests during walking experience the space interest "visit garden — read garden — enjoy garden". Green hill and beautiful lotus in the hallway, the verses in the lobby and koi swimming in the hall, all together outline the rich and elegant life of the south. Third,the archaic Chinese rhymes of salting. Reversed with the sweet taste of southern cuisine, the food caters the local people. Fourth, Yangtze river and canal. Without the Yangtze river and the canal, there is no prosperity of Huaiyang cuisine. Green hill and beautiful lotus in the hallway and wall paintings of picturesque Yangtze write the prosperity of the South.

Based on the color extraction of white walls and dark gray roof tiles, designers join modern fashionable elements such as camel,buff and achromatic color, making the clean and elegant space more intimate and modern.There are contracted furniture instead of official hat chairs, to match with the elegant scene. The overall design makes the tradition and moderness form a blend of the emotion, and the guests dining in this space have rather different feelings.

井河公馆
私家菜
RIVER RESTAURANT

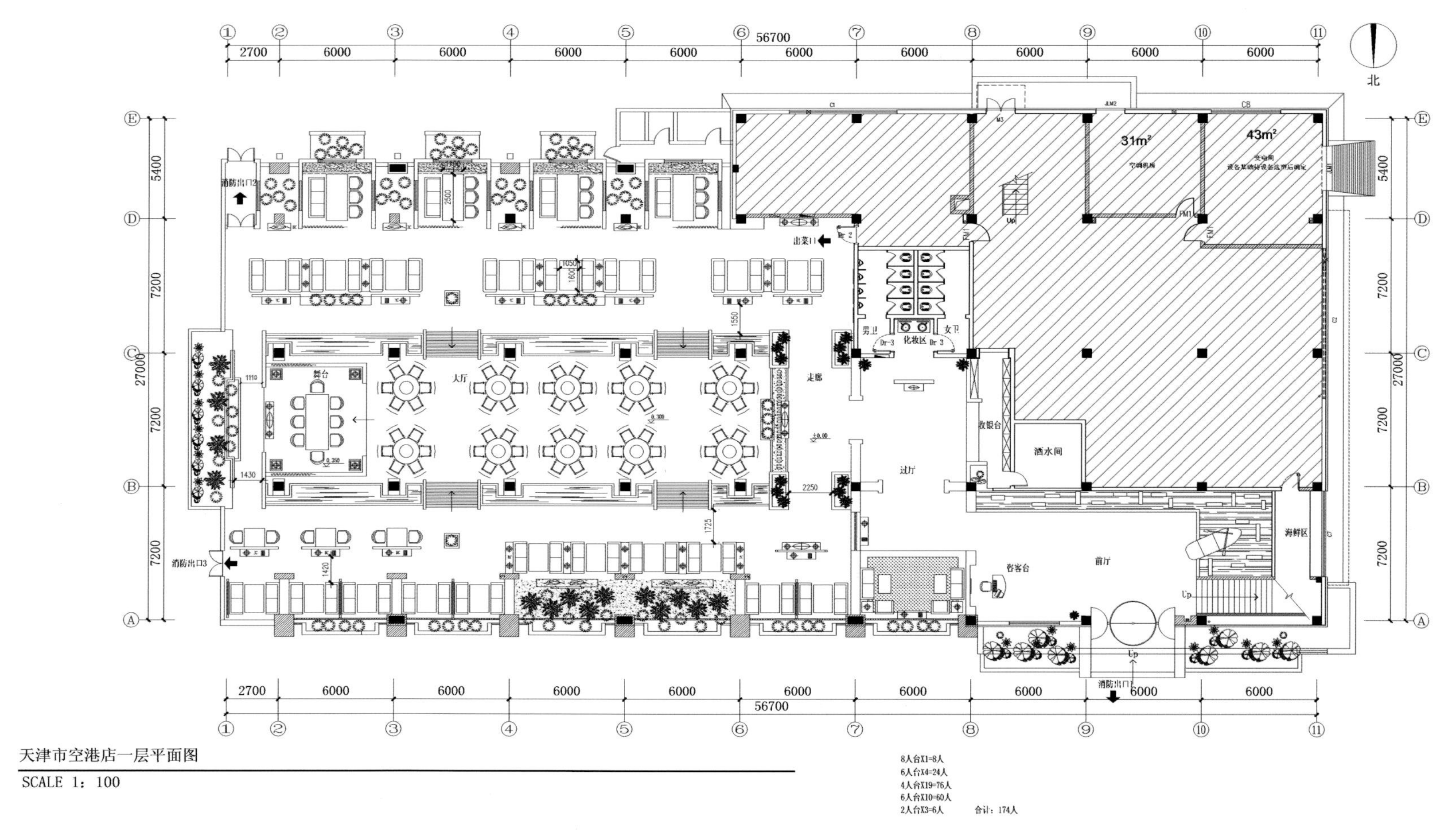

天津市空港店一层平面图

SCALE 1：100

盐商文化是贯穿设计的主线，以此展开设计思路。扬州人多以盐务为生，习于浮华，精于肴馔，故扬州筵席全国驰名。由盐商和盐官所形成的饮食规范是一种场面浩大、环境典雅、菜肴奇特、选料精严、食器精美的饮食风格。

主线中的节点设计：一是淮扬菜的精髓。以味为核心，以养为目的，延展到空间设计中"以景为核心，以惬意为目的"虚境实景相结合（水墨江南墙画、小桥流水虚实交映）。二是盐商富而雅的生活情趣。生活化的空间"静""雅""趣"透过景观的布置，使得客人在行走过程中体验"游园—读园—赏园"的空间情趣。门厅的青山碧荷，过厅的诗词歌赋，

大厅的锦鲤戏水共同勾勒出江南富雅的生活。三是腌渍古韵。一反江南菜的甜腻，适合天津当地口味。四是长江与运河。没有长江与运河，就没有淮扬菜的繁荣，门厅的青山碧荷，墙壁淡雅的江南主题的泼墨画书写着江南的一派繁荣。

在提取粉墙黛瓦的色调基础上，加入现代时尚的元素，无彩色加驼色米色系列使得清雅的空间多了分亲切与现代感。家具没有采用传统的官帽椅而是简约的现代家具，从色彩关系与造型上与儒雅的氛围相呼应，使得传统与现代形成一种情感的交融，客人在此空间就餐别有一番感受。

30

20

10

Fu Lin De Seafood Hot Pot Restaurant in Beijing

北京福临德海鲜火锅

Design Agency: Dashidai Design & Consulting Co., Ltd. **Designer:** Wu Xiaowen, Zhang Yinghui **Area:** 3500 m^2 **Completed Date:** 2013, 4
Photography: Xing Zhentao **Materials:** White jazz, Wood stone, Leather, Metal grille

设计单位：大石代设计咨询有限公司 设计师：吴晓温、张迎辉 面积：3500 m^2 完工时间：2013 年 4 月 摄影：邢振涛 主要材料：爵士白、木纹石、皮革、金属花格

Festive elements opened the story of Fu Lin De culture, with art glass lamp sea showing decorating joy atmosphere. The office of the main background is the interpretation of the Chinese character "Fu", expounds the creative source of blessing culture.

Rooms emphasize the sense of architecture, hexahedron modeling is contracted, with copper bar outline makes them comprehensive rich sense of pictures. The design focuses on the elaboration of furniture, to create a simple and elegant temperament. Underground bulk hall, Maitreya Buddha sculpture means "happiness is very simple ".

大厅以大红灯笼高高挂的喜庆元素拉开福临德文化的故事序幕，以艺术玻璃灯海展现张灯结彩的喜悦氛围。过厅主背景则是福字演绎，阐述了福文化的创意来源。

包间强调内建筑感，六面体造型简约，以铜条勾勒使得包间富有画面感，设计的重点则放在对家具陈设的推敲上，营造简约而精致的气质。地下散厅，弥勒佛的雕塑寓意着“幸福很简单”。

福临德
海鲜火锅
私人医生

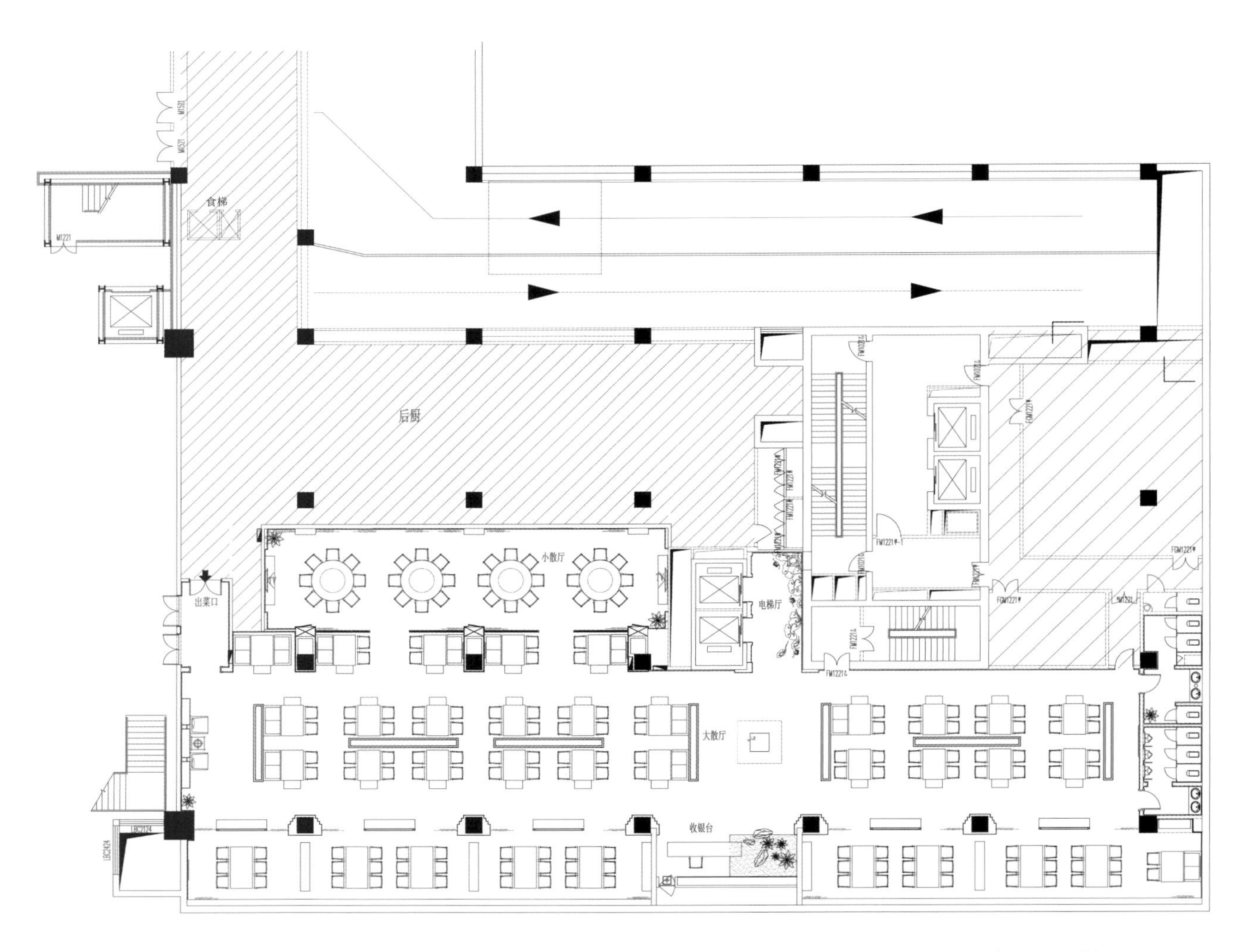

地下一层平面图

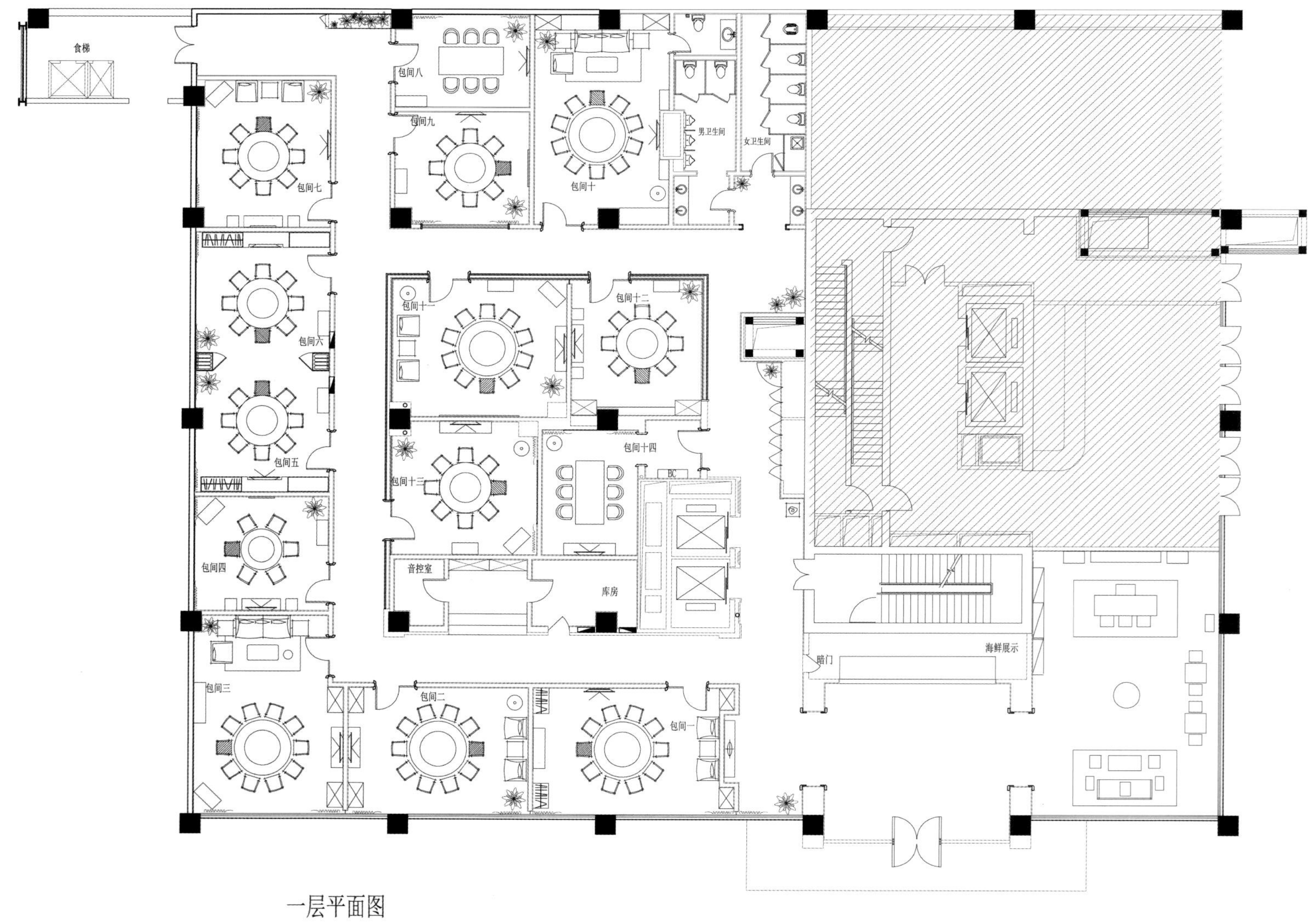
食梯
包间八
包间九
包间十
男卫生间
女卫生间
包间七
包间六
包间五
包间四
包间三
包间二
包间一
包间十一
包间十二
包间十三
包间十四
音控室
库房
暗门
海鲜展示

一层平面图

安全出口
EXIT

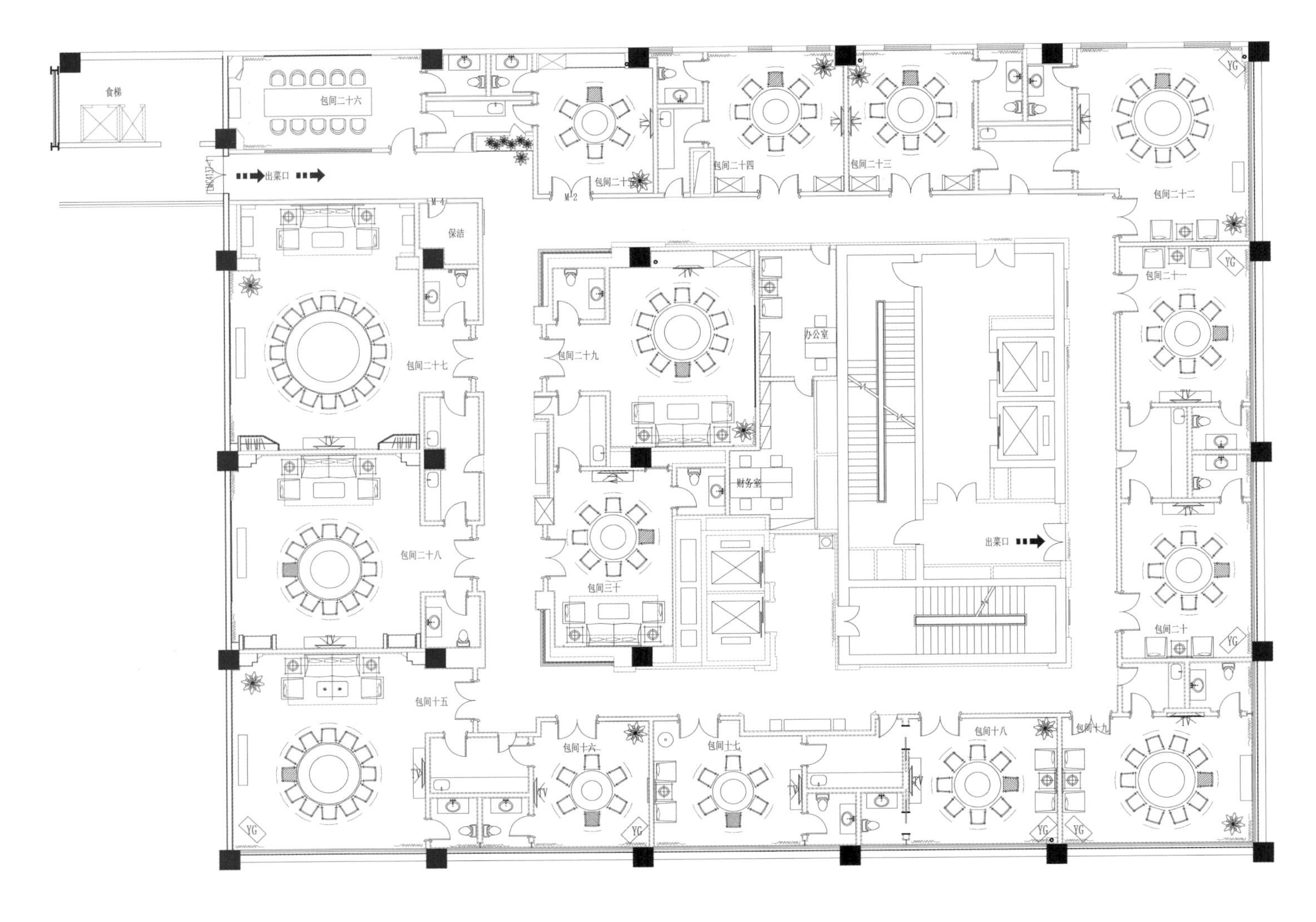
食梯
包间二十六
出菜口
包间二十四
包间二十三
包间二十二
保洁
包间二十一
办公室
包间二十七
包间二十九
财务室
包间二十八
包间三十
出菜口
包间二十
包间十五
包间十六
包间十七
包间十八
包间十九

二层平面图

安全出口
风华正茂
寿比南山
海纳百川
喜从天降
五福和合
信守不渝

Yueyu Restaurant in Guangzhou

广州悦语酒家

Design Agency: Shenzhen Hua Space Design Consulting Co., Ltd. **Designer:** Xiong Huayang, Xu Wenwen **Area:** 2000 m^2 **Materials:** Latex paint, Marble, Wood veneer, Mosaic, Wallpaper, Carpet

设计单位：深圳市华空间设计顾问有限公司 设计师：熊华阳、徐雯雯 面积：2000 m^2 主要材料：乳胶漆、大理石、木饰面板、马赛克、墙纸、地毯

Yueyu Restaurant mainly operates Cantonese cuisine, targets on gatherings, wedding and business. The professional restaurant designers believe that the restaurant should firstly focus on the uniform of food and brand image. Secondly, it should view from the position of the consumers' perspective.

According to its market positioning, the designers used modern fashion style combined with Southern culture to convey a brand story and cultural connotation of the dining space. Entering into the restaurant, you can see the lobby at first glance. Lobby's design not only shows the brand image, but also plays a role of shunting crowds in the dinning peak. According to the functional requirements of the restaurant management, the half of area is open space, and the other is suits.

The corridor plays a supporting role, which expresses the southern taste. Atrium designed as a semi-open suit form which can reduce the noise of the restaurant space. When the restaurant creates a quiet atmosphere, many consumers feel comfortable. Restaurant toilets are another place to reflect the restaurant positioning grade and management level. The toilets in Yueyu focus on function — a large area of the mirror material expands the visual area and the gray and black marble material is easy to clean .

In generally speaking, restaurant design should focus on uniform of brand and image. Brand and image must be consistent with the brand positioning, in order to reflect the brand to convey ideas, value or meaning.

V10

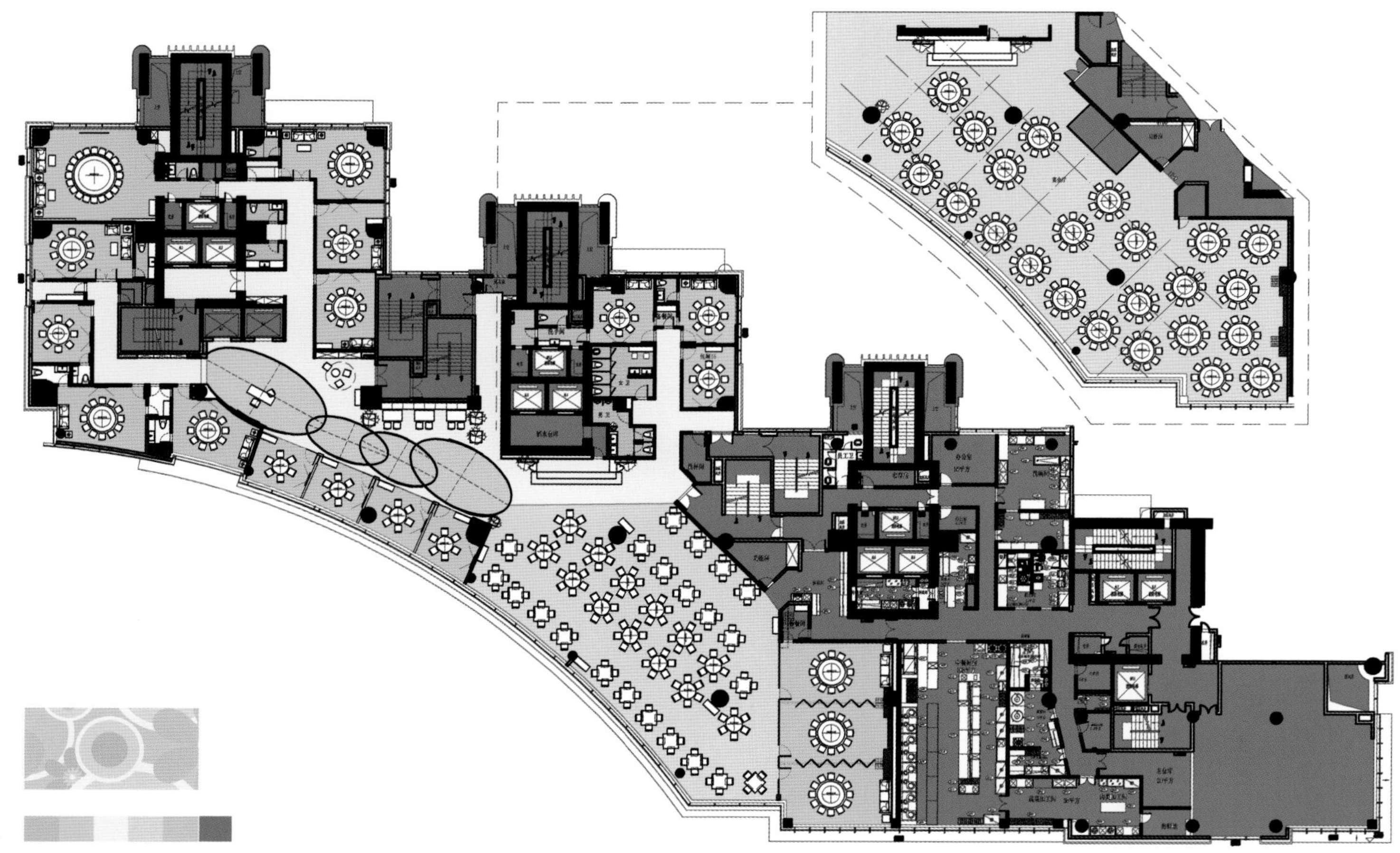

悦语餐厅以经营粤菜为主，目标消费群体是朋友聚餐、家庭聚会、喜宴及商务宴请。专业餐饮设计师认为，首先要注重餐饮品牌的形象整体统一，其次要从消费者的立场与角度出发。

通过餐厅的市场定位，设计师在现代时尚风格的基础上，结合岭南文化设计元素融合其中，传达一种有品牌故事，有文化内涵的餐饮空间。进入餐厅首先要看到的是前厅，前厅设计不仅直观地向客户展示品牌形象，更是在用餐高峰时起到分流人群或等候的作用。根据餐厅经营的功能性需求，其中一半面积为开放式大堂空间，一半面积规划为包房。

餐厅的走廊空间起到一种承上启下的辅助作用，将空间打造成具有岭南蕴味的温馨空间。中庭位置使用隔断将空间形成半包间的形式，分区隔断减少餐厅空间的嘈杂，当餐饮空间营造出这种安静的氛围，给消费者的感觉会舒适很多。餐厅洗手间是另一个体现餐厅定位档次及管理水平的地方，悦语餐厅洗手间设计注重功能使用，大面积使用镜面材料，起到放大空间的视觉效应，而且灰黑色的大理石材料容易清洁。

总体来说，餐厅设计中要注重品牌形象的统一，品牌形象的载体就是要与品牌定位相一致，才能体现出品牌所传达的思想、价值或内涵。

109

135

135
128

Color Hot Pot — Sanqin Original Hot Pot

唐锅——三秦原创火锅

Design Agency: Dashidai Design & Consulting Co., Ltd. **Designer:** Wu Xiaowen **Area:** 1000 m² **Completed Date:** 2012.2 **Photography:** Wu Xiaowen
Materials: Stone(Athens beige), Floor tiles, Wallpaper, Stainless steel, Mercury mirror

设计单位：大石代设计咨询有限公司 设计师：吴晓温 面积：1000 m² 完工时间：2012 年 2 月 摄影：吴晓温 主要材料：石材（雅典米黄）、地砖、壁纸、不锈钢、水银镜

居鍋
三秦原创火锅
hotpot
COLOR
居鍋
colorhotpot
三秦原创火锅
假日浴场

Tang Guo — Sanqin original hot pot, pursues a kind of low-pitched and restrained taste, low and light life style, in the cultural orientation and the environment, lighting, music, services. Flying apsaras, peony, Datang red build elegant space of light life, without too much warm reception, push the door to see the long corridor of Datang red. The people who are smiling to welcome guests are ceramic maids of Tang Dynasty (tri-coloured glazed pottery of the Tang Dynasty). And with light environment, the slow pace of carding mood, guests are relishing the ancient culture of Tang Dynasty.

Change disadvantages into features. This project is located in a remote place. Business district is behind commercial body along the street, and guests need going through the narrow and long lane to enter into the business hall. The height of the building is less than 2.6 meters. So, it is very hard to build the "Tang Dynasty" in space. "A thing is valued if it is rare", and this is also some kind of scarcity, so use "light" to shape the "prosperous" of Datang. The original architectural appearance is glass curtain wall (bath forms), which is refined into red hollow curtain device through flying ribbon, showing Datang red and flying culture — "light

and prosperous". Long and narrow corridor is designed as art gallery"Datang red" , and along the street to enter, there is much expectation and relaxation, so we foreshadow behind the gallery (make design changes in two dimensional space of the hall), and open the door to see lively and fashionable space, as waterscape, oil painting of the Tang Dynasty and traditional maid's accessory decorate cultural expressions of the space. We give guests more segments of memory in small space, (thin is light, and less is expensive) and leave appreciators of space more recall of Datang. The corridor of Catering area, rooms and lobby, respectively interspersed with the flying apsaras, peony, maids of Tang Dynasty, chinaware, build the impress of Datang history. Dining chair's metal framework, ceiling of mercury mirror, black gloss tile on the ground, by reflecting and controlling the brightness of space make the space extended. The concept" thin is light, and less is expensive" solves the problems that rise in building flourishing age in the narrow and short space.

唐锅——三秦原创火锅，从文化定位及环境、灯光、音乐、服务都追求一种低调的、内敛的品位和慢的、轻的生活方式。飞天、牡丹、大唐红营造轻生活的雅致空间，没有过分的热情接待，推门而入是悠长的大唐红长廊，迎接客人的是仪容端庄、笑容可掬的陶瓷唐朝仕女（唐三彩），轻的环境、慢的脚步梳理心情，回味着久远的大唐文化。

变劣势为特色，本项目坐落在偏僻之处，营业区在沿街商业体后方，需穿过狭长的胡同才能进入营业大厅，建筑层高不足 2.6 米。要营造大唐盛世可见空间难度非同一般，“物以稀为贵”，这也算“稀缺”吧，于是用“轻”去塑造大唐的“盛”。原有建筑外观为玻璃幕墙（浴池业态），透过飞天飘带提炼为红色镂空幕墙装置，体现大唐红与飞天文化——“轻而盛”。狭长的走廊设计为“大唐红”艺术长廊，沿街而入多了份期待与放松，于是我们在长廊之后埋下伏笔（门厅的二度空间做设计转变），推门而入映入

眼帘的便是明快时尚的空间感受，水景、大唐盛世油画、仕女饰品装置着空间的文化表情，小空间当中我们留给客户更多的片段记忆，（疏则轻少则贵）留给空间的品赏者更多的对大唐的追忆。餐饮区走廊、包间、过厅分别穿插了飞天、牡丹、唐仕女、瓷器等陈设品来营造大唐的历史印记，餐椅金属框架、吊顶水银镜、地面黑色高光地砖，通过反射以及控制空间亮度使得空间得以延展。“疏则轻少则贵”的理念，解决了狭窄空间转化为盛世空间的问题。

青岛啤酒 NBA

302

清
香

Pine Crane Building

松鹤楼

Design Agency: Shanghai Vjian Design Office **Designer:** Song Weijian, Yu Wanbin **Location:** Suzhou **Area:** 3800 m^2
Materials: Wang tiles, Clip silk glass, Wood stone

设计单位：上海微建建筑空间设计事务所 设计师：宋微建、于万斌 地点：苏州 面积：3800 m^2 主要材料：望砖、夹绢玻璃、木纹石

Pine Crane Building is designed as the orientation of "new Jiangnan style", which is using modern design idea, modern technology to recombine Gusu classicism. Song Weijian gained inspiration from Suzhou Shantang Street reconstruction, and played use of it in the design of "old dongwu" restaurant, becoming more pure and refined till today.

松鹤楼的设计，是将其定位为"新江南风格"，也就是采用现代设计理念、现代工艺将姑苏古典风格重新组合而成。宋微建从苏州山塘街改造中获得灵感，并在"老东吴"食府的设计中充分发挥，延续至今，愈发纯粹与精练。

Feast in Tonight

今夜宴语

Design Agency: Yijing Interior Design Agency in Chengdu **Designer:** Zhong Qiming, Huang Ying **Area:** 1300 m^2 **Location:** Yibin, Sichuan
Client: Fulin Jincui Concept Catering Company **Completed Date:** 2011 **Photography:** Wang Feng

设计单位：成都易景室内设计事务所 设计师：钟其明、黄莺 面积：1300 m^2 地点：四川宜宾 客户：福林锦翠概念餐饮公司 完工时间：2011 年 摄影：王峰

Restaurant appearance design is concise and easy. Restaurant is located on the second floor. The middle of the ground insets a large dark black gold Afghani flower which has a strong guiding role; Metope seagull cream-colored stone with Mosaic black mirror steel is more visual wallop, and solved the vision problems that the first floor to the second floor stair looks very narrow.

Entrance of the second floor rail used tower top bead type column, novel and generous, rendering sequence aesthetic feeling, and solves the space to surround close the topic; Rest area corresponding to the reception hall and background wall painting flowers design, make stairwells and rest activity zoning.

The left side of reception hall is the installation art landscape pool, pond wall uses mirror material to embody the "plum culture" Junlian, bright colors and light mottled ball in reflection make people happy, and become a theme in the design of the window.

餐厅外观设计简洁大方，餐厅位于二楼，地面中间镶嵌大面积深色阿富汗黑金花，起到很强的引导作用；墙面海鸥米黄横石材之间镶嵌黑镜钢，更具视觉冲击力，并且解决了一楼到二楼楼梯间十分狭长的视觉问题。

二楼入口栏杆采用到顶塔珠式立柱，新颖大方，呈现序列美感，并且解决了空间围合问题；接待大厅与大厅休息区相对应，喷绘花卉图案背景墙，使楼梯间与休息区动静分区。

福林錦翠
歡迎光臨
Welcome
歡迎光臨
Welcome

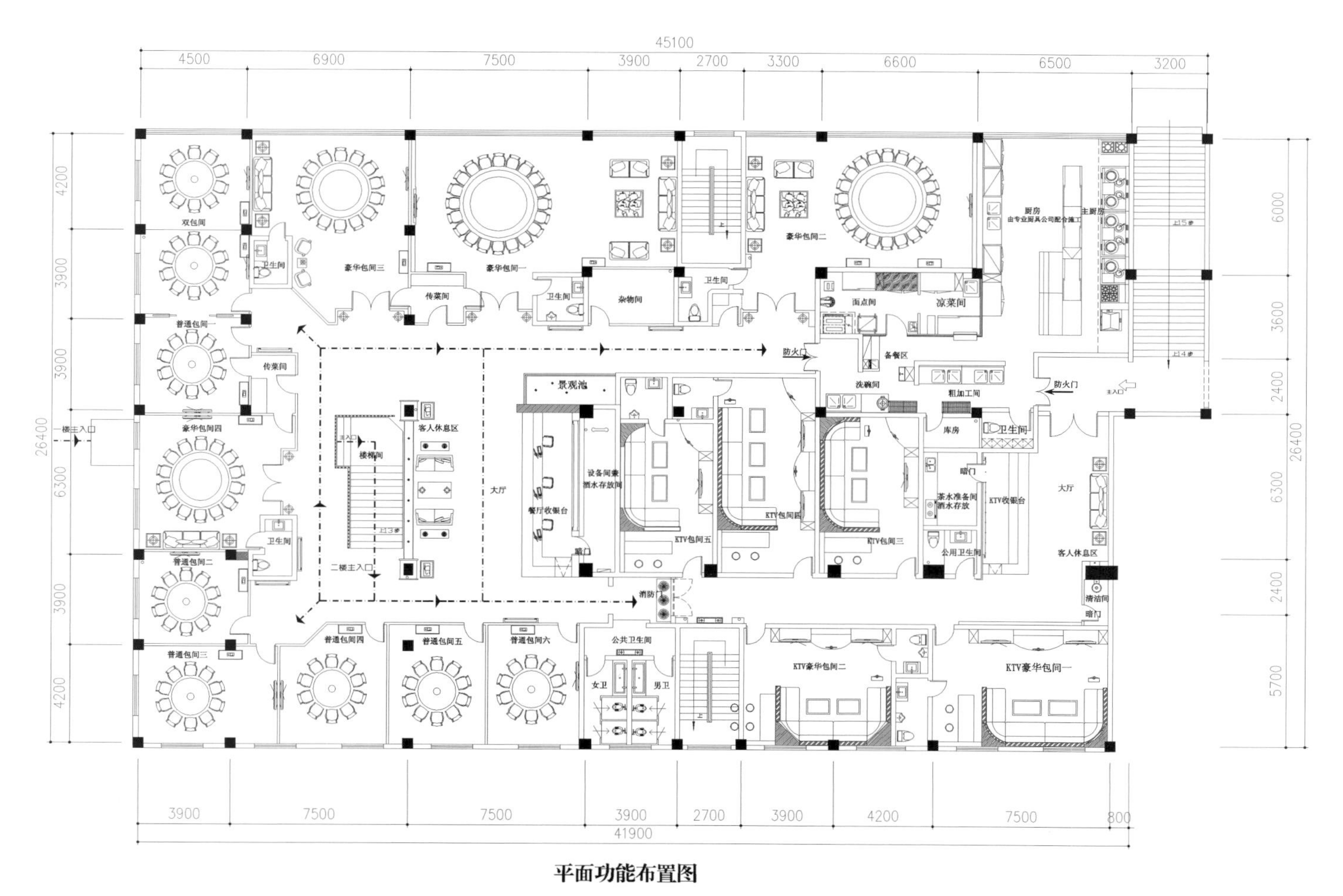

平面功能布置图

接待大厅左侧为装置艺术景观池，池内壁墙面、顶面采用镜面材质，地面铺以白沙衬底，彩色装饰水果球，正体现了筠连本土的“李子文化”，鲜艳的色彩与光影斑驳的玻璃球交相辉映，使人就餐心情愉悦，并成为设计中的主题亮点。

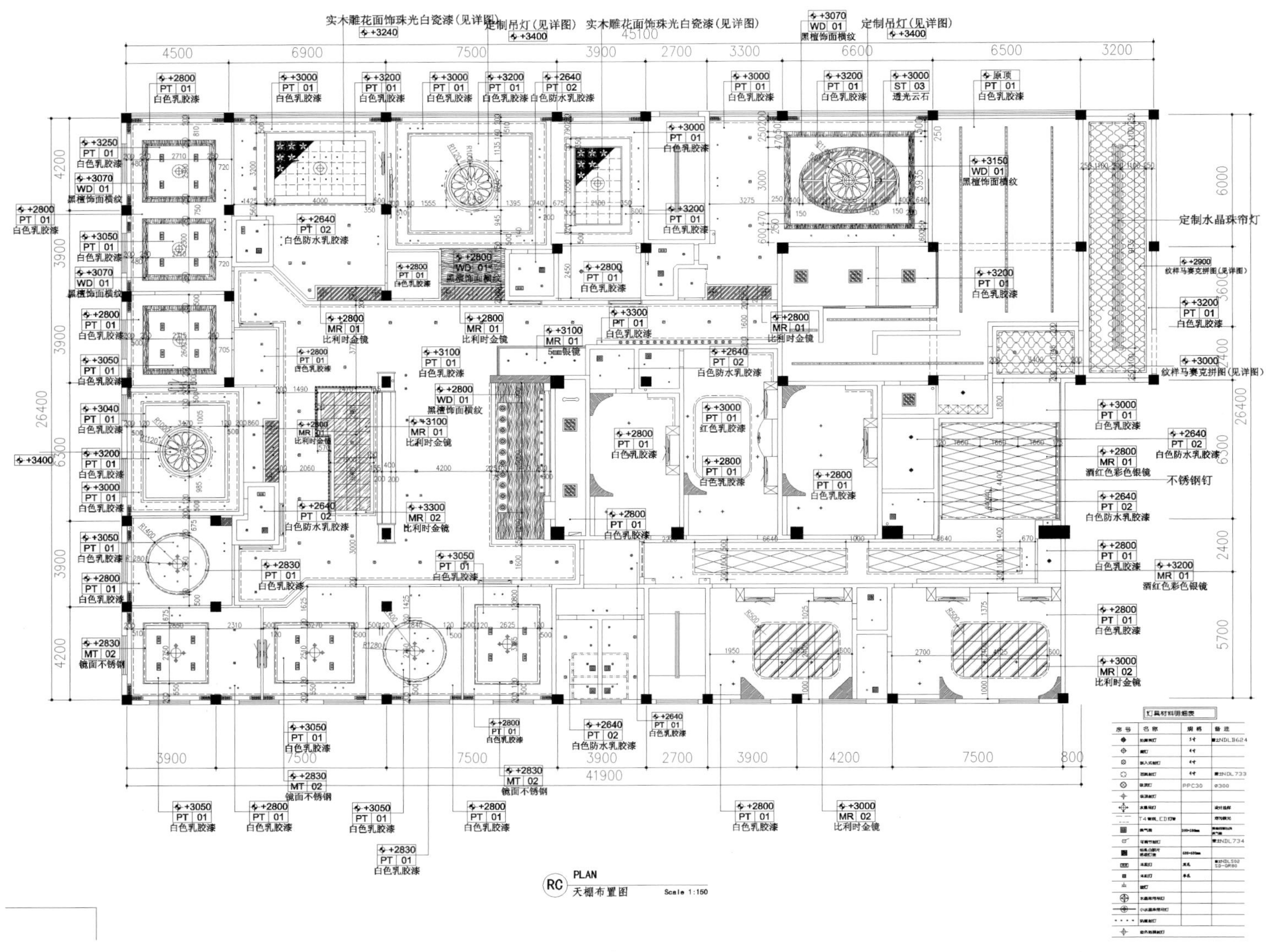
RC PLAN
天棚布置图
Scale 1:150

伍

Country Yard at Tijiao Park

堤角公园农家小院

Design Agency: Wuhan ONE ZERO Design Co., Ltd. **Area:** 2300 m^2 **Location:** Tijiao Park in Wuhan **Completed Date:** 2011.1 **Photography:** Cao Jun

设计单位：武汉壹零空间设计有限公司 面积：2300 m^2 地点：武汉堤角公园 完工时间：2011 年 1 月 摄影：曹军

As one of the most famous chain restaurant in Wuhan, "Country Yard" was redesigned according to the environment of Tijiao Park. It can meet the requirement of overall environment in style and form!

作为武汉本土一家深得百姓熟悉的"农家小院"连锁店之一，本案在结合堤角公园实际的主题环境下，进行了重新定位，使得本案在设计风格及表现形式上更加地符合整体环境的需要！

棋牌烟酒

祖德宗功師範長
天地君親師
道冠古今
德參天地
天高地厚君恩重
馬到成功

樂
康
壽
特许专供店
Licence for special shop
馬到成功

洗手间
TOILET

灭火器箱
火警119

鲜榨果汁
阴米肚片汤
灭火器箱
火警119

硚口

ROCHO

六潮

Design Agency: DOLONG Design **Area:** 1200 m^2 **Completed Date:** 2012.4 **Materials:** Fair-faced concrete, Brick, Floor, Art steel net
Photography: Jinxiao Space Photography

设计单位：董龙设计 面积：1200 m^2 完工时间：2012 年 4 月 主要材料：清水混凝土、金砖、地板、艺术钢网 摄影：金啸空间摄影

ROCHO is a cultural fashion agency which combined restaurant and bar.

六潮餐饮，是一家集创意菜和酒吧为一体的文化时尚机构。

大潮
ROCHO
創意中國菜
鷄尾酒吧

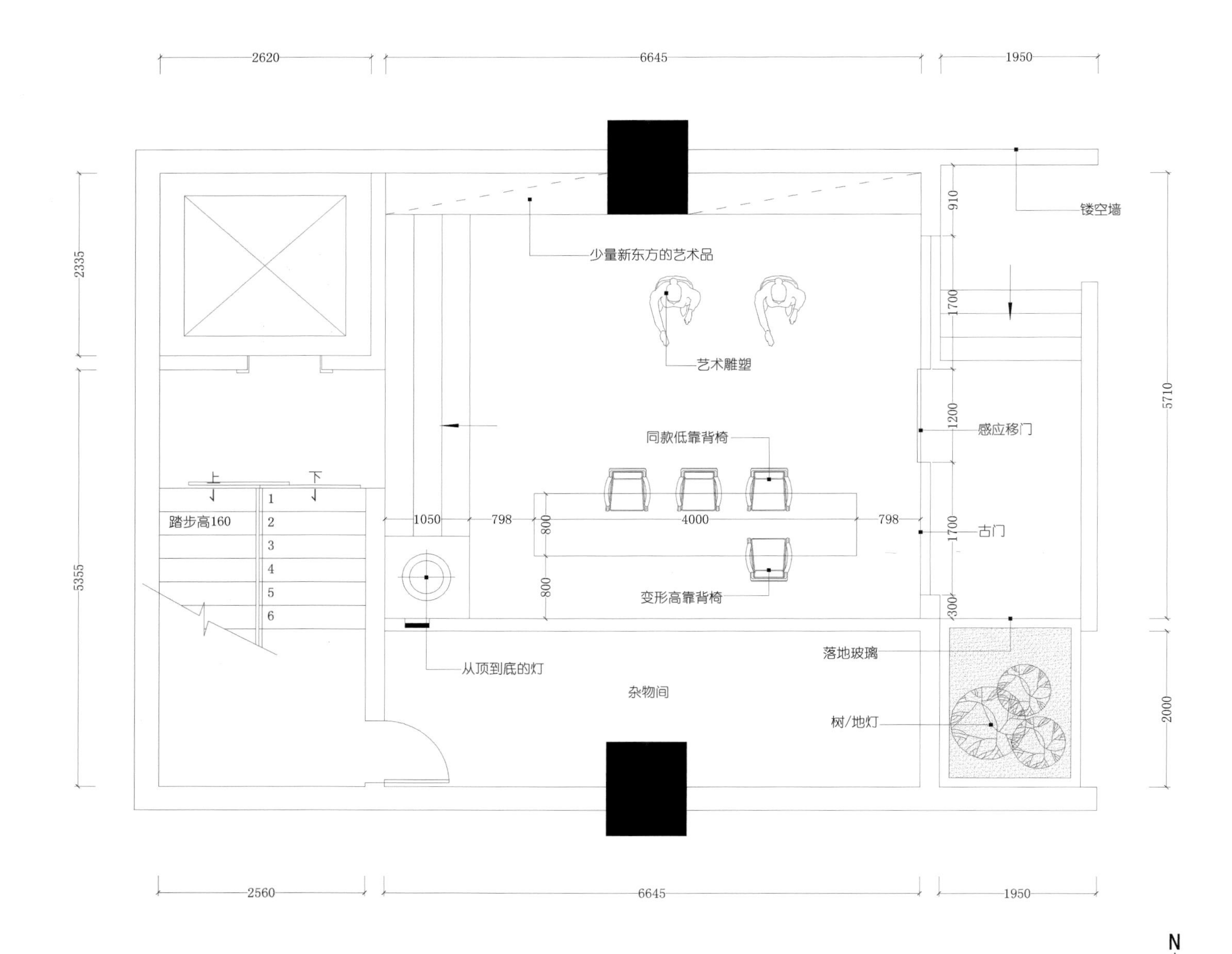
2620
6645
1950
镂空墙
少量新东方的艺术品
艺术雕塑
感应移门
同款低靠背椅
上
下
踏步高160
1
2
3
4
5
6
1050
798
4000
798
800
800
古门
变形高靠背椅
从顶到底的灯
杂物间
落地玻璃
树/地灯
910
1700
1200
1700
300
2335
5355
5710
2000
2560
6645
1950

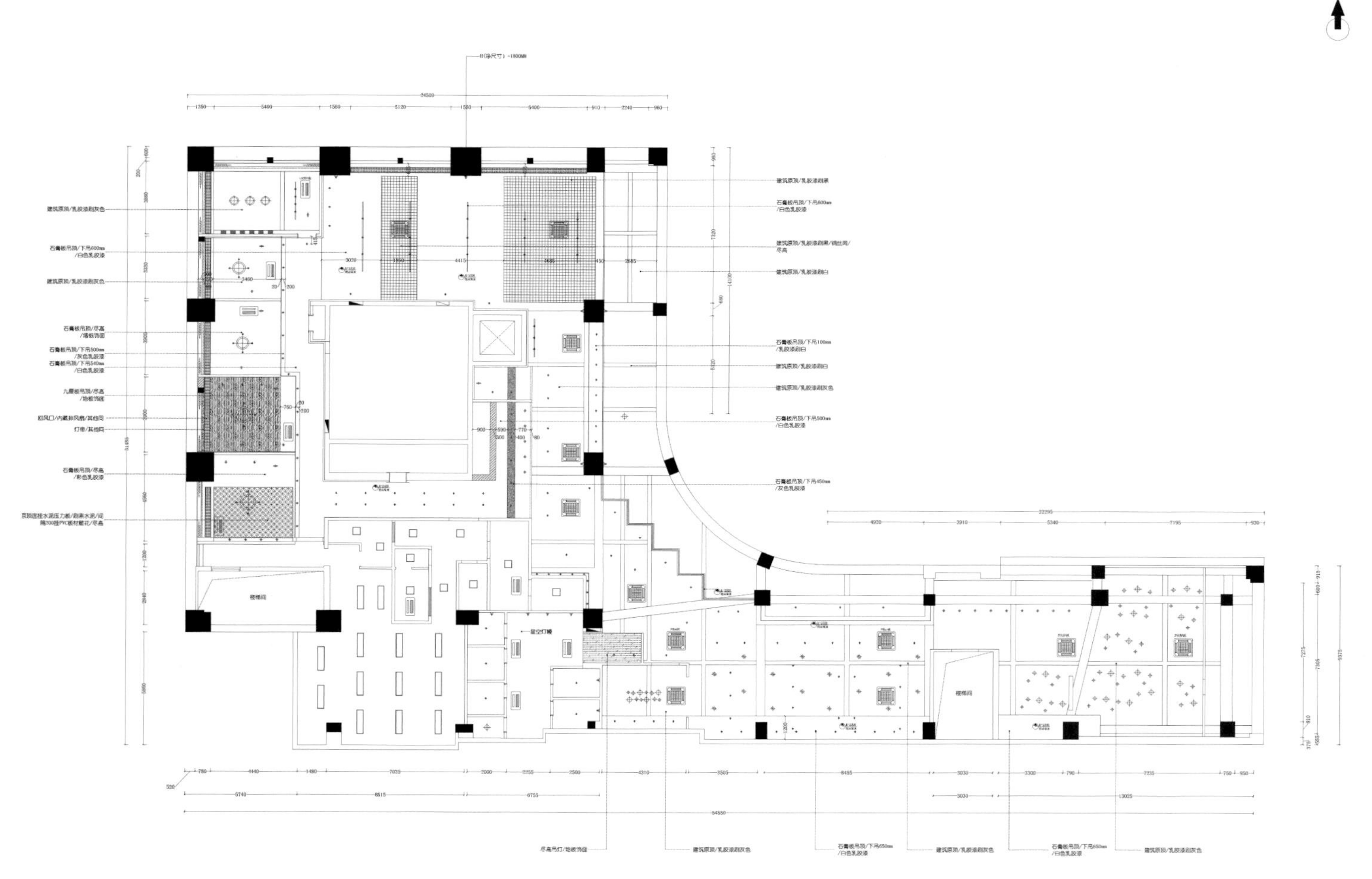
N

B
C
A
D

Ballantine's

Old Dongwu Restaurant in Yadu, Suzhou

苏州老东吴食府雅都店

Design Agency: Shanghai Vjian Design Office **Designer:** Song Weijian, Zhang Nan **Area:** 1500 m^2 **Location:** Suzhou, China
Client: Old Dongwu Restaurant in Suzhou **Cooperation:** Huayuan Decoration **Materials:** Wood, Bamboo, Wood stone, Wall brick

设计单位：上海微建建筑空间设计事务所 设计师：宋微建、张楠 面积：1500 m^2 地点：中国苏州 客户：苏州老东吴食府 合作单位：华远装饰
主要材料：实木条、竹子、木纹石、城墙砖

This project is characterized to Suzhou cuisine. Four years ago, the VIP rooms on second floor were redesigned . The overall spatial layout quoted Suzhou garden, creating a double-deck feeling, and won a good reputation in the industry.

The hall focuses on nature, emphasizing the interaction between man and environment, and making the space more varied. Corridor and room was separated by louvered wooden, which not only skillfully ensures certain privacy, but does not completely block the external view, creating a harmonious, dynamic dining space for customers. Material application is in compliance with the design concept of simple and natural. The traditional elements appear careful and natural. It is embodied with the graceful beauty of classic culture however without the heaviness of deliberation.

本项目是一家以经营本地家常菜为特色的苏州餐馆。四年前，对该餐馆的二层贵宾包厢区域进行了重新设计，整体空间布局上引用苏州园林的布局，营造出楼中楼的感觉，在业界赢得不错的口碑。

一层大厅的设计更加着重自然，强调人与环境的互动体验，使空间更富于变化。廊和厢通过百叶式的木质隔断，既巧妙地保证了一定的私隐性，又没有完全阻隔外界的视野，为顾客营造一个和谐、畅动的用餐空间。材料应用上也遵从朴实、自然的设计理念。而散落在各处的传统元素细心而不觉做作，让人随时领略文化经典的雅致之美，却没有那种刻意为之的沉重感。

金桂

Wu Families Restaurant with Theme of Red Chamber

吴地人家红楼主题餐厅

Design Agency: Shanghai Vjian Design Office **Area:** 1700 m^2 **Location:** Beijing **Materials:** Strip ceramic brick, Wooden lattice, Marble mosaic

设计单位：上海微建建筑空间设计事务所 面积：**1700m²** 地点：北京 主要材料：条形陶砖、木花格、大理石拼花

Dream of the Red Chamber is a story of Jinlin in the south of the Yangtze River. Designers adopt the pattern of the south garden in the layout, and apply the skills such as "burst to screen", "prefect to repress" in the design. When the space is for the stage, people become the characters. With the twelve beauties of Jinglin with red golden-line on setting wall of main hall, fan-shaped shakedown on the ground of the entrance, round portal on both sides, stone bridge indoor, once coming in you are brought into the scene of the story. Though the whole space is only 1700 square meters, it builds the effect of different landscape of each step, a scene of one room. Service staff of the dining room often joke "I'm lost".

On space turn, or use big white light boxes before which flower scene is decorated, or use hollow long window of beautiful cases which gives the effect of partition and continuousness. So, it doesn't produce dead angle, and at the same time, let people feel the continuation of the space. Part of boxes are designed by the concept of pavilion, and pavement of lotus pond and the form of dismantling the bridge, effectively solve the repeatedly dull feeling the dense boxes are easy to form.

A large area of black glass bricks are used on the wall of public part replacing of sheathing tiles that are often used in the design of garden, which is more suitable for the title, and makes the whole space more contemporary. There is ceramic mosaic of 1.6 meters ornamented on the ground of black material, enriching the spatial diversity and richness.

《红楼梦》是一部发生在江南金陵的故事。在平面布局上采用江南园林式的格局，”欲则屏 ”，“嘉则收”等手法应用其内。空间为舞台，人即成为了主角，正厅背景墙上红底金线的十二金钗，入口地面扇形的地铺，两侧的圆洞门，门内的石桥，使顾客一入门即被带入了故事的情景之中，整个空间虽只有 1700 平方米，但却营造了一个移步换景，一景一空间的效果。餐厅服务人员也常戏言“我迷路了”。

在空间转折上或采用大的白色灯箱，灯箱前布置花景，或采用镂空花格长窗，却起到了隔而不断的效果。既不会产生死角，同时也让人感觉到了空间上的延续。部分包厢

以亭子的概念设计，以及地面花池、折桥形式的铺装，都有效地解决了密集包厢容易形成的重复单调感。

公共部分的墙面上采用了大面积的黑色玻璃条砖，代替园林式空间设计上常用到的望砖，既入题又使整个空间更加富有现代感。黑色材质的地面上点缀有直径 1.6 的瓷片拼花，都更加充实了空间的多样性、丰富性。

稻香
525

Haidilao Hot Pot Restaurant

海底捞

Design Agency: Wisdom Space Design **Area:** 1891 m^2 **Location:** Suzhou **Completed Date:** 2012.8 **Photography:** Zhu Shenfen
Materials: Mirror stainless steel, Suede fireproof board, Toughened paint glass, Printing film, Stone, Square tube matte paint spraying, Yarn clamping glass, Latex paint

设计单位：睿智汇设计 面积：1891 m^2 地点：苏州市 完工时间：2012 年 8 月 摄影：朱沈锋 主要材料：镜面不锈钢、绒面防火板、钢化烤漆玻璃、喷绘软膜、人造洞石、方管哑光漆喷涂、夹纱玻璃、乳胶漆

Haidilao is a Chinese famous hotpot brand. This project is the second branch in Suzhou, located in Suzhou Wuzhong Unity South Bridge. It was designed by Wisdom Space Design . Design scheme satisfies the terminal consumer's aesthetic taste and rigorous comprehensive functional requirements, at the same time, in the continuation of traditional culture, embodying era's enthusiasm, design elements and methods promote the haidilao brand spirit, successfully consolidate the haidilao brand business competitiveness.

海底捞——中国著名火锅品牌，本案为海底捞的苏州第二分店，坐落于苏州市吴中团结桥南，空间设计委托睿智汇设计团队倾力打造。设计方案满足了终端消费者的审美趣味和严谨周详的功能需求，在延续传统文化的同时，融入了时代的热情，设计元素与手法发扬了海底捞的品牌精神，成功巩固了海底捞品牌的商业竞争力。

收银台
海底捞

海底捞
海底捞

海底捞

海底捞

HaiDiLaoHuoGuo
海底捞火锅
自选水果台

067

HaiDiLaoHuoGuo
海底捞火锅
自选小料台

Jin Mama Hand-made Noodle Restaurant

晋妈妈手擀面王府井店

Design Agency: LJ Design Office in Shanghai **Designer:** Zhang Guanglei **Area:** 150 m^2 **Location:** Zibo, Shandong **Photography:** Yu Jian

设计单位：上海光磊（LJ）设计师事务所 设计师：张光磊 面积：150 m^2 地点：山东淄博 摄影：于健

"Jin Mama" is pleased to spread the Chinese food culture. It not only fully inherited the essence of traditional hand-made noodle, but also uses modern cooking technology to change the traditional fast food. From this concept, Jin Mama Hand-made Noodle Restaurant in Zibo takes the new Chinese style design to show the simple Chinese classical taste.

"晋妈妈"以传播中华面食文化为己任，既要充分继承传统手擀面的精华，又要用现代烹饪技术改变传统快餐的随意性。从这个思路出发，晋妈妈手擀面淄博王府井店，采用新中式设计风格，品味简约中式的古典味道。

可口可乐
Jin Mama
晋妈妈

JinMama
晋妈妈

福
可口可乐
JinMama
晋妈妈

Jindama
晋妈妈

吉祥如意

吉祥如意
请勿吸烟

Wumizhou Restaurant

五米粥中润店

Design Agency: LJ Design Office in Shanghai **Designer:** Zhang Guanglei **Area:** 200 m² **Location:** Zibo, Shandong **Client:** Sun Fadong
Completed Date: 2012.12

设计单位：上海光磊（LJ）设计师事务所 设计师：张光磊 面积：200 m² 地点：山东淄博 客户：孙发东 完工时间：2012 年 12 月

This project is a chain Chinese food institution which is mainly with seafood hot pot. The dishes are unique and the design style is close to modern Chinese style. Limited in size, it maximums in space utilize. Many corners are used as the spices tables. Design technique is concise, and combined with a small amount of Chinese elements. The black is simple and calm.

本案是一家以海鲜火锅为主的连锁中餐机构，菜品独具特色，设计风格为现代中式，受面积限制，在空间利用上尽量做到餐位数最大化，很多边角利用起来做调料台备餐橱等。设计手法力求简练，并融入少量中式元素。色调以黑色为主，简练又不失稳重。

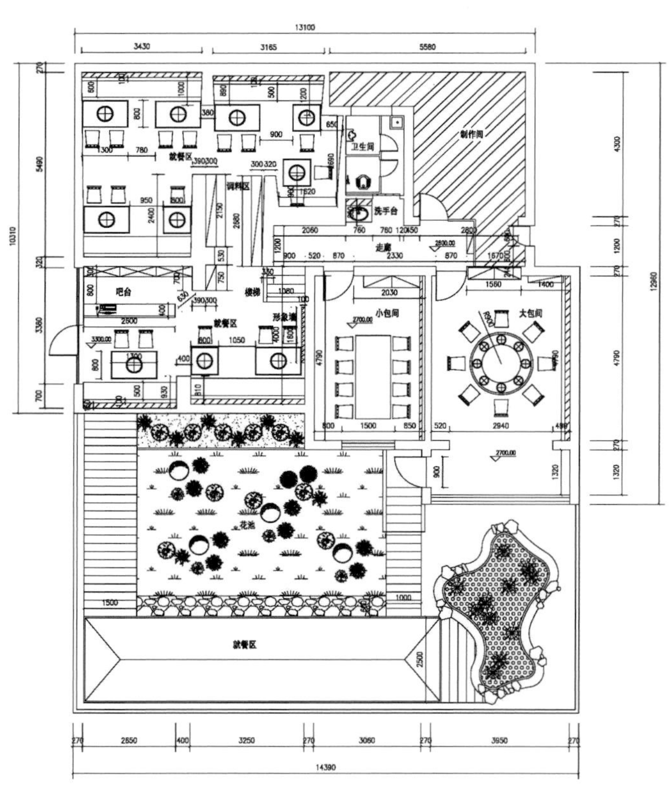

五米粥中润店平面布置图 1:100

Jin's Food Restaurant in Xi'an

西安市锦食记

Design Agency: SHENZHEN YIDING DESIGN & DECORATION CO., LTD.   **Designer:** Wang Kun   **Area:** 250 m^2 **Location:** Xi'an, Shanxi

设计单位：深圳市艺鼎装饰设计有限公司　设计师：王锟　面积：250 m^2　地点：陕西西安

There are some swaying lotuses dancing in the pool. The successful application breaks pattern of boring, and brings some fresh air to the indoor.

Jin Food Restaurant has clear functional areas which is perfect with light and color. The white bar divided the restaurant into two pieces. Rather than any other color but white is in order to echo the lotus background color , balance tables and chairs and light color.

随风摇曳的荷花在池面起舞。设计师成功地运用这幅画打破方正格局的沉闷之感，给室内带来几许清新空气。

锦食记餐厅的功能分区明确，光线和色彩的处理恰到好处。白色横栏的运用清晰地把格局划分成两块，简单明了。选用白色，而不是其他色，则是与荷花背景色呼应，平衡桌椅和灯光的色彩。

收银处
A—01

CASHIER

Renben

CASHIER
Rubin

Wuji Old Pan Coast City Shop

吴记老锅底海岸城店

Design Agency: SHENZHEN YIDING DESIGN & DECORATION CO., LTD. **Designer:** Wang Kun **Area:** 602 m^2 **Location:** Shenzhen

设计单位：深圳市艺鼎装饰设计有限公司 设计师：王锟 面积：602 m^2 地点：深圳市

The space with fresh and quiet traditional aroma without losing nicety, makes every guests that come in feel enjoyable and comfortable, which is the ultimate pursuit of this case.

On the material selection of the whole space, we use the wood veneer to decorate all sides, and at the same time, march with materials such as leather, cloth art and so on, to make the space more stable and dignified under light rendering. And the join of elements of traditional ink painting adds the surrounding environment some elegance and beauty, which is traditional inheritance, and also the traditional promotion. On the ceiling, the design of lamp has also become a bright spot in the space, which is a combination of modern geometry and postmodern ideas.

Whether the integral space outside or VIP high-end sideboard, everywhere manifests a kind of both Chinese and western and orderly dining environment, which is also comfortable, not low and deep.

在清新、淡雅的传统韵味中，又不缺现代的精细，让每一个进入这个空间的客人，既感到一份惬意，又有一份舒适是本案的最终追求。

在整个空间的材料选择上，我们采用了木饰面布置四周，同时，配以皮革、布艺等材料加以协调，使空间在灯光的渲染下更加沉稳、端庄。而传统水墨画等元素的加入，使得周围的环境里增添了几分文雅、儒秀的内容，这是对传统的继承，亦是对传统的发扬。天花顶上灯的设计也成为空间里的亮点，是现代几何和后现代创意的结合。

不论是外面的整体空间，还是 VIP 的高备餐柜，处处体现出一种中西兼备而有序，氛围舒适而不低沉的就餐环境。

A18

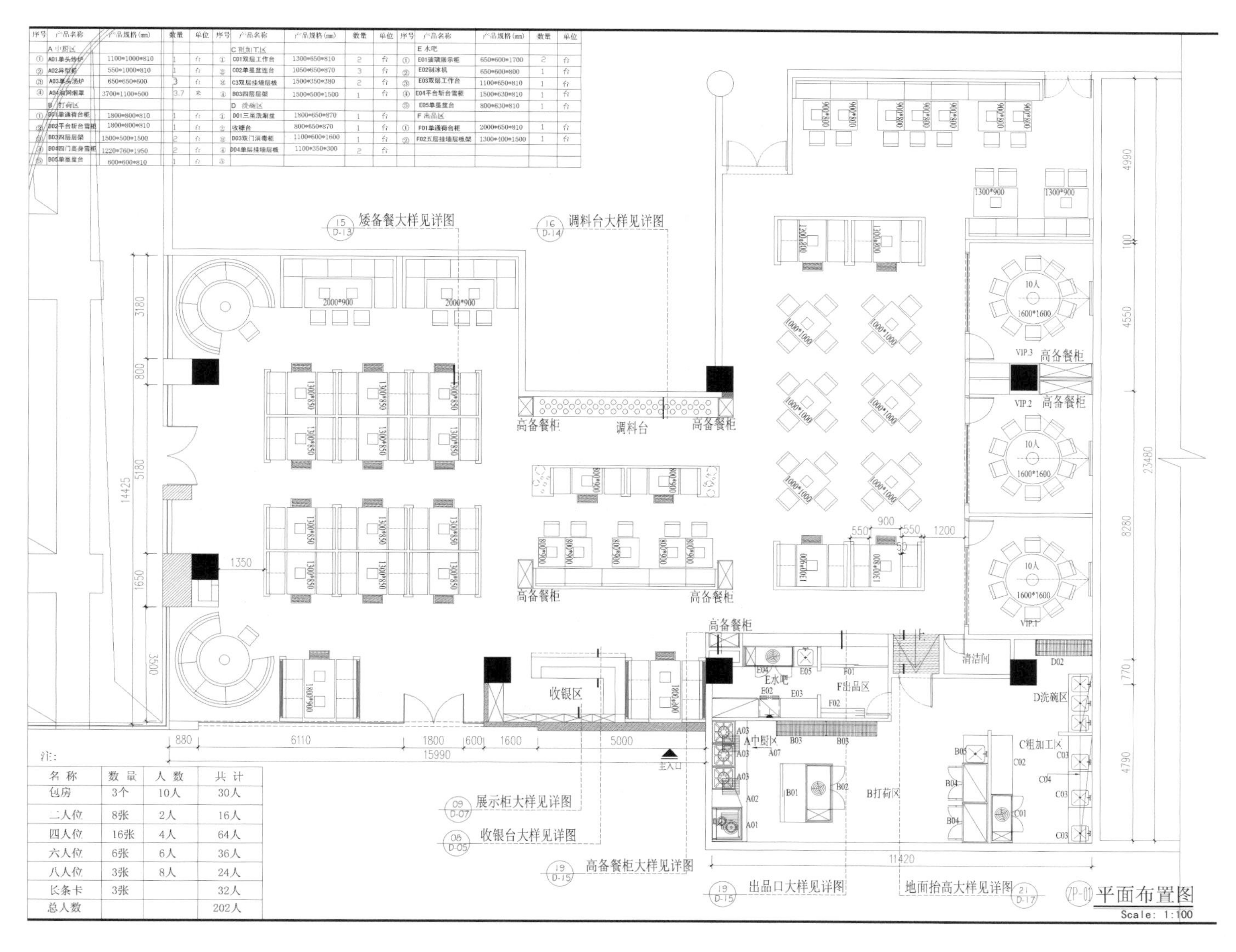

序号	产品名称	产品规格(mm)	数量	单位	序号	产品名称	产品规格(mm)	数量	单位	序号	产品名称	产品规格(mm)	数量	单位
	A 中厨区					C 粗加工区					E 水吧			
①	A01单头炒炉	1100*1000*810	1	台	①	C01双层工作台	1300*650*810	2	台	①	E01玻璃展示柜	650*600*1700	2	台
②	A02异型柜	550*1000*810	1	台	②	C02单星盆连台	1050*650*870	3	台	②	E02制冰机	650*600*800	1	台
③	A03单头汤炉	650*650*600	3	台	③	C3双层挂墙层板	1500*350*380	2	台	③	E03双层工作台	1100*650*810	1	台
④	A04油网烟罩	3700*1100*500	3.7	米	④	B03四层层架	1500*500*1500	1	台	④	E04平台斩台雪柜	1500*630*810	1	台
	B 打荷区					D 洗碗区				⑤	E05单星盆台	800*630*810	1	台
①	B01单通荷台柜	1800*800*810	1	台	①	D01三星洗涮盆	1800*650*870	1	台		F 出品区			
②	B02平台斩台雪柜	1800*800*810	1	台	②	收银台	800*650*870	1	台	①	F01单通荷台柜	2000*650*810	1	台
③	B03四层层架	1500*500*1500	2	台	③	D03双门消毒柜	1100*600*1600	1	台	②	F02五层挂墙层板架	1300*400*1500	1	台
④	B04四门高身雪柜	1220*760*1950	2	台	④	D04单层挂墙层板	1100*350*300	2	台					
⑤	B05单星盆台	600*600*810	1	台	⑤									

注:

名称	数量	人数	共计
包房	3个	10人	30人
二人位	8张	2人	16人
四人位	16张	4人	64人
六人位	6张	6人	36人
八人位	3张	8人	24人
长条卡	3张		32人
总人数			202人

B10

B05

安全出口
EXIT

老鍋底 麻辣火鍋
海岸城店
8月18日盛大开幕
二层大中庭
顾客服务台
时尚展示区
国际酒吧街
sCat
sCat
A20

Contributors 设计师名录

Cai Zongzhi

蔡宗志

Cai Zongzhi, born in Taibei, graduated from Architect Department of Danjing University, gained a diploma from Architect DESA, Paris,France Ecole special d'architecture in 1997, set up Atelier de L'A.V.I.S in 2005.

蔡宗志，台湾台北人，台北淡江大学建筑系毕业，1995 年法国留学，1997 年获得法国巴黎建筑专业学院建筑师文凭（Architect DESA ,Paris,France Ecole special d' architecture），2005 年成立法惟思设计工作室（Atelier de L' A.V.I.S）。

Chen Yonggen

陈永根

Chen Yonggen graduated from the indoor environment art design major in 1999, held a post of manager of Design Department in Hangzhou Jiuding Decoration Company in 2004, and held a post of Design Director in Ningbo Longding Decoration Company in 2008.

Now, he is the design director and the originator of SOKU ART DECORATION DESIGN CO., LTD. in Ningbo, as well as senior member and senior interior designer of IAI Asia Pacific Association of Designers. He has won 2010 biennial Asia-Pacific interior design grand award competition for best newcomer, and earned award nomination of catering in 2011 Asia-Pacific green elite global tournament. His works covers catering, hotel, club,villa,etc.

陈永根，1999 年毕业于环境艺术室内设计专业，2004 年于杭州九鼎装饰任设计部经理，2008 年于宁波龙顶装饰任设计总监。

宁波斯库装饰设计有限公司任设计总监，宁波斯库装饰设计有限公司创始人，IAI 亚太设计师联盟资深会员、资深室内设计师。曾获 2010 亚太室内设计双年大奖赛新人奖，2011 亚太绿色全球精英邀请赛餐饮类大奖提名。作品涵盖餐饮、酒店、会所、别墅等。

DING HE DESIGN

河南鼎合建筑装饰设计工程有限公司

DING HE DESIGN, in the form of combination of large design engineering company, operating area of 2000 m^2. Main undertake indoor and outdoor architectural decoration design and construction of project, project in hotel, restaurant, club house, office, real estate projects is given priority to. The company has excellent design team, which includes 2 China hundreds of excellent interior architects, 4 senior interior architects, more than 50 interior designers, with the design concept of innovation strives for realism, the construction of scientific and standardized management, the service of integrity and high quality completed multiple large-scale project design, construction, and won a good reputation in the industry.

河南鼎合建筑装饰设计工程有限公司是以强强联合形式组成的大型设计工程公司，经营面积约 2000 m^2。主要承接室内外建筑装饰设计及施工工程，项目以酒店、餐饮、会所、办公、地产类项目为主。公司拥有卓越的设计团队，其中中国百名优秀室内建筑师 2 人、高级室内建筑师 4 人、室内设计师 50 余人，以创新求实的设计理念，科学规范的施工管理，优质诚信的服务高质量地完成了多个大型设计、施工项目，并在业界赢得了良好的口碑。

董龙设计

DOLONG Design

董龙设计是一所专业高端设计公司，成立于 2008 年 5 月 4 日。品牌理念为：创造精品，让实景作品诠释实力。品牌规模：董龙设计、大品专业施工、董龙陈设、设计施工均是乙级资质。我们的经营模式：设计 + 基础装修，设计 + 基础装修 + 主材选择。经营范围：室内空间设计、公共空间设计、独立软装饰设计。

董龙设计一直致力于完整家居课题的思考和探索，设计作品曾多次获全国性专业比赛各大奖项，作品也曾被各大出版社定期约稿作为专业性书籍出售。

DOLONG Design, established on 4th, May, 2008, is a high-end professional design company. Design concept: create excellent project, and make the real project express our strong strength. Brand status: DOLONG Design, high quality professional construction, DOLONG display, design and construction, are all having Grade B qualification. Our business model: design+foundation decoration, design+foundation+advocate material selection. Business scope: interior space design, public space design and independent soft decoration design.

DOLONG Design advocates itself to think and explore the full household topic, and its projects have won many famous awards in the national professional completion, as well as published in professional architecture books.

东航装饰室内设计建筑有限公司（万维空间设计机构）

Donghang Decoration InteriorDesign Construction Co., Ltd. (Wanwei Space Design)

东航装饰室内设计建筑有限公司（万维空间设计机构），2006 年由江西知名设计师高波、邹巍先生创建于中国首个被国际授予国际“陶瓷之都”的城市——景德镇。

历经 8 年的潜心探索、发展与沉淀，始终坚持从“以人为中心，由美出发”，坚信“大境空间止于想象”，不断拓展产品线，涉及建筑景观规划创意设计、室内建筑创意设计、陈设艺术创意设计和美学文化创意设计，东航装饰已成为江西室内建筑设计行业高端品牌。

The design agency of Donghang Decoration Interior Design Construction Co., Ltd. (Wanwei Space Design), is founded by famous designers of Gao Bo and Zou Wei of Jiangxi in 2006, and it is located in Jingde Town which is the first city awarded the international “ceramic capital” in China.

For eight years of exploration, development and precipitation, the company have always adhered to the concept"people-centered, starting from beauty", firmly believed that "great space design is decided by imagination", and continuously expanded the product line, involving architectural landscape planning and creative design, indoor architectural creative design,display art creative design and aesthetic culture creative design. Now, the company has become the high-end brand in the interior architectural design industry in Jiangxi province.

蒋国兴

Jonny

蒋国兴，1996 年毕业于厦门工艺美术学院。现为苏州叙品设计装饰工程有限公司总经理兼设计总监，是中国室内设计协会会员。作品曾多次发表。

Jonny, graduated from Xiamen Arts and Design College in 1996. He is general manager and design director of Xu Pin Design and Decoration Engineering Co., Ltd. in Suzhou. He is a member of China Interior Design Association. And his works have been published several times.

JOY-CHOU YI INTERIOR DESIGN STUDIO

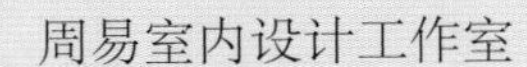

周易室内设计工作室

JOY-CHOU YI

1959 Born in Taichung, Taiwan, R.O.C.

1979 Study architecture and the interior design independently

1989 Establish JOY INTERIOR DESIGN STUDIO

1995 Establish JOY CONCEPT ARCHITECT STUDIO

周易

1959 出生于台中市

1979 以自学方式学习建筑与室内设计

1989 创设 JOY 室内设计工作室

1995 创设 JOY 概念建筑工作室

Li Zhou Li Gan

李舟 李干

Designers Li Zhou and Li Gan are brothers in design field. Both of them are the hundreds of excellent interior architects, form L + B Chongqing Interior Design Studio, now named as Zhumu Decoration Design Co., Ltd. in Chongqing. They specialize in providing unique creative design concept; their interior design has the personality; and they have a research in a variety of commercial space, the fashion restaurant, theater design and real estate in the clubhouse, sales offices, decoration design, and they have high visibility in the design industry.

设计师李舟、李干是室内设计界一对知名的兄弟组合，两人均为全国百名优秀室内建筑师，组建重庆 L+B 室内设计工作室，现更名重庆筑木装饰设计有限公司，擅长于提供独特的有创意的设计构思，做具有个性的室内设计，对各种商业空间、时尚餐厅、剧场设计及房地产的会所、售楼处、装饰设计有研究，在业内有较高知名度。

Li Ming

李明

Li Ming, interior architect, design director of Zhihua Decoration in Jinan. In 2006, established Ming Design Agency. Now, design director in Bright Design.

Master of Shandong Art College, professor of Shandong University.

Vice Secretary-general in IDA, secretary-general in Shandong Design Agency, member of China's Architecture Agency.

李明，高级室内建筑师，曾担任济南志华装饰工程有限责任公司设计总监，2006 年创立“明”设计顾问机构，现任 Bright Design 设计总监。

山东艺术学院硕士研究生，现为山东大学威海分校艺术学院的兼职教授。

IDA 国际设计师协会济南分会副秘书长，山东设计委员会副秘书长，中国建筑装饰协会会员。

朗昇国际商业设计有限公司

Lonson International Commercial Design Co., Ltd.

朗昇国际商业设计有限公司是一家专业从事室内空间设计与环境艺术设计的顾问公司，于 2003 年创立于深圳。公司成立伊始便志存高远，引进香港及国际先进的设计行业管理模式，倡导原创设计，致力于打造中国高端商业空间、酒店、娱乐空间、餐饮空间、样板房、售楼处、会所、高档豪宅家居、高端写字楼等室内设计精品，为客户提供全程式设计顾问服务。

Lonson International Commercial Design Co., Ltd., established in Shenzhen in 2003, is a consultant company specialized in interior design and environment art design. The firm is ambitious from the very beginning by introducing the Hong Kong and international advanced management model for design industry. It advocates the original design and devotes to create the high quality design projects among the fields of commercial space, hotel, entertainment space, catering, sample house, sales office, club, villa, high-end office, providing the clients with one-stop consultancy service.

石炎森

Shi Yansen

石炎森，毕业于西安美术学院环境艺术设计系。2005 年就职于城市人家装饰主任设计师。2010 就职于上海憬华设计工程有限公司担任设计总监。2011 年至今，担任西安森鹏室内设计工程有限公司的设计总监。设计感悟：创意成就设计，设计感悟人生。

Shi Yansen, graduated from Xi 'an academy of fine arts. In 2005, worked as a director in City Family Decoration ; In 2010 work as a design director in Shanghai Jinghua Design Engineering Co.,Ltd. as a director. Since 2011, as the design director of Senpeng Interior Design Co., Ltd. in Xi'an. Design inspiration: creative achievements in design, design comprehension in life.

宋微建

Song Weijian

宋微建，出生于 1959 年，自幼酷爱绘画，深圳大学建筑系毕业。早年在深圳蜻蜓设计从事工业设计。20 世纪 80 年代开始从事建筑室内设计，在餐厅、会所、酒店、博物馆、老建筑改造设计等方面见长。目前为中国建筑学会室内设计分会副理事长，上海微建建筑空间设计事务所首席设计师。2011 年及 2012 年连续被中国十家建筑室内专业媒体评为“中国室内设计十大影响力人物”之一。崇尚自然、崇尚“天人合一”的中国宇宙观，在设计中对中国传统文化进行深入探索与发展，是构建宋微建空间设计的核心观念。近年来创作了一系列具有“新江南形式语言”特色、影响深远的空间作品和原创产品。

Song Weijian was born in 1959 and graduated from Architecture Department of Shenzhen University. He was fascinated in drawing when he was very young. Song was engaged in industry design in the early years of his career. Then in 1980s, he moved into the field of interior design, now specializing in the design of restaurant, club, hotel, museum, and building reconstruction. He is the deputy director of China Institute of Interior Design (CIID), the chief designer of Shanghai Vjian Design Office. He was evaluated by 10 architecture & interior design related media as the top ten influential interior designers in China in consecutive years of 2011 and 2012. Holding the world view of advocating nature and the combination of nature and humanity, he has a deep research and development of Chinese traditional culture in his design, and this is also the core concept for Vjian Design. Recent years he was witnessed the creation of a series of space projects and original products with the characteristics of “new Jiangnan style language” which generates a profound influence.

The Gazer Design Center

杭州肯思装饰设计事务所

The Gazer Design Center was established in 2003 in Hangzhou, China. We provide a full range of quality and creative interior design, planning and project management services for villa, chain-stores, hotels, offices, clubs, exhibitions and other business spaces. Our group devotes to create the best design for clients.

杭州肯思装饰设计事务所成立于 2003 年，公司设于中国杭州，致力于提供全方位、高品质、富于创造性的室内设计和项目规划执行服务，业务范畴涵盖别墅、专卖店、酒店、办公室、会所、展览等商业空间。我们的团队旨在以最好的设计回报客户。

Wang Kun

王锟

Wang Kun, design director of SHENZHEN YIDING DESIGN & DECORATION CO., LTD., director of the Shenzhen Designers Association, member of China Building Decoration Association, member of Chinese decoration.

王锟，深圳市艺鼎装饰设计有限公司设计总监，深圳设计师协会理事，中国建筑装饰协会会员，中国装饰协会会员。

Wisdom Space Design

睿智汇设计

Wang Junqin, as the head of Wisdom Space Design, is a famous interior designer in Taiwan, the international leader of entertainment space design. He has won many national and international awards for his flexible and suitable designs and has been acclaimed as a designer possessing both sense and sensibility, establishing his status in China's design field. The Wisdom Space Design company he founded is famous for its entertainment space design in China's design field, named one of China's top ten most influential catering and entertainment design institutions, and China's most valuable enterprise of interior design, having serviced numerous well-known brands and enterprises. The concept of being prudent, relentless and calm in design is his attitude towards design as well the motivation of his entire team.

王俊钦，睿智汇设计公司掌门人，台湾著名设计师，国际新娱乐空间设计大师。他因地制宜的设计屡获国内外大奖，被誉为理性与感性双全的设计师，在中国设计领域享负盛名。他所创立的睿智汇设计以娱乐空间设计着称于中国设计界，被评为中国十大最具影响力餐饮娱乐类设计机构、中国最具价值室内设计企业等殊荣，服务于众多的知名品牌企业。“慎言笃行，精筑致远”是他对设计的态度，也是整个睿智汇团队的动力。

武汉壹零空间设计有限公司

Wuhan ONE ZERO Design Co., Ltd.

武汉壹零空间设计有限公司是一家具有先锋设计理念、顶尖设计人才的室内设计公司，以高品质的设计水准进驻人们的生活，升华个人生活品味。我们拥有华中地区最强的设计团队，具有丰富的室内家装、公装设计经验；专业服务住宅平层、复式、别墅、房地产样板房、售楼部、商铺、咖啡馆、会所等设计。团队以大胆创新、超前的设计风格和高品质生活定位为不懈追求的目标，并以此为信念，解决在设计中遇到的一切挑战；精诚服务、共筑未来，团队自信、轻松、诚恳的工作态度服务于对设计高要求的客户群体。团队提供室内硬装设计、建材产品、软装家具饰品整体设计陪购等服务。

Wuhan ONE ZERO Design Co., Ltd. is a interior design company having pioneer design idea, top design talent, with high quality design standards in people's life, the sublimation of personal life taste. We have the strongest design team in central China, has the rich interior decoration and design experience; Professional services for residential flat layer, double entry example room, villa, real estate, sales department, shops, cafes, clubs, etc. Bold innovation, advanced design style and positioning for the relentless pursuit of high quality life goals to solve all the challenges encountered in the design; Sincere service, altogether building the future, the team confidence, honest work attitude in the service of high requirements for the design of customer group. Teams provide indoor hard outfit design, building materials products, purchasing soft outfit furniture accessories overall design etc.

熊华阳

Xiong Huayang

熊华阳，深圳市华空间设计顾问有限公司总经理、设计总监，高级室内建筑师，中国建筑学会室内设计分会会员，CIID 中国建筑学会室内设计分会深圳（第三）专业委员会委员。

Xiong Huayang, the general manager and design director of Shenzhen Hua Space Design Consulting Co., Ltd., the senior interior designer, the member of China Institute of Interior Design (CIID) and the committee member of Shenzhen (the Third) Professional Committee of CIID.

许建国

Xu Jianguo

许建国，CIID 中国建筑学会室内设计分会会员，国家注册高级室内建筑师。担任中国建筑室内环境艺术专业高级讲师，另为中国美术家协会合肥分会会员。安徽许建国建筑室内装饰设计有限公司设计主持。

Xu Jianguo, CIID China Institute of Interior Design branch member, national registered senior architect. As China Building indoor environment art professional senior lecturer, another for member of the Chinese Artists Association in Hefei. Design host in Anhui Xu Jianguo Architectural Interior Decoration Design Co., Ltd.

Yiduan Shanghai Interior Design

上海亿端室内设计

The founder of Yiduan Shanghai Interior Design, Xu Xujun (chief executive creative director), is the director of east china branch of International Association of Architecture and Interior Design, director of International registered Artists and Designers Association, International senior resigistered interior designer , a member of the Association of Chinese Interior Designer, instructor of the fourth National Colleges Space Design contest and the front page figure of 2012-2013 China Interior Deigner.

上海亿端室内设计创办人徐旭俊（首席执行创意总监）是国际建筑装饰室内设计协会华东分会理事、国际注册艺术家设计师协会理事、国际注册高级室内设计师、中国室内设计师协会 / 专业会员、第四届全国高校空间设计大赛四大高校实战导师和 2012—2013 中国室内设计师年度封面人物。

Zhang Guanglei

张光磊

Zhang Guanglei, was born in February1980. Now lives in Zibo, Shandong, member of China Building Decoration Association, registered senior architect, Shandong University of Technology external lecturer, Since 2003 he engaged in interior design, and has worked in the Han Rui International (Taiwan), sun membrane structure (Japanese), HZS (US) design company. In 2009 he established LJ Design Office in Shanghai.

张光磊，1980 年 2 月生，现居山东淄博，中国建筑装饰协会会员，注册高级室内建筑师，山东理工大学外聘讲师，2003 年从事室内设计至今，先后任职于韩瑞国际（台资），太阳膜结构（日资），HZS（美资）等设计公司。2009 年成立上海光磊（LJ）设计师事务所。

Zhang Yingjun

张迎军

Zhang Yingjun, the member of International Federation of Interior Designers/Architects, the director of Chinaroot Design, the top 100 outstanding youth interior architect, member of the 23rd Professional Committee under China Institute of Interior Design (CIID), member of Design Committee under China Building Decoration Association, the vice president of China Ji Xiaolan Seminar, the vice president of Hebei Restaurant Industry Association and the general designer of Dashidai Design & Consulting Co., Ltd.

Projects: Huang Tea House, Macao Doulao, Yuewei Restaurant etc.

张迎军，IFI 国际室内建筑师联盟设计师，大木设计中国理事，中国百人杰出青年室内建筑师，中国室内建筑学会室内设计分会第二十三届专业委员会委员，中国装饰协会设计委员会委员，中国纪晓岚研究会副会长，河北省饭店烹饪行业协会副会长，大石代设计咨询有限公司总设计师。

代表作品：凰茶会、澳门豆捞、阅微食府、东北虎黑土印象、无名居等。

钟其明 Zhong Qiming

钟其明，成都易景室内设计事务所设计总监。黄莺，成都易景室内设计事务所首席设计师。成都易景室内设计事务所主力为室内建筑空间设计，由知名设计师担纲主持设计。自成立以来参与了国内外多项大型室内外设计项目——餐饮、酒店、金融、办公、百货、娱乐、会所、展场、别墅、房地产、样板间等多项高端设计项目，并在多项国内及国际设计大赛中获得殊荣，作品被多家国内核心媒体、网站、杂志、学术刊物连续刊载并全国发行，受到客户广泛好评。并且现在也致力于高端设计的前期研发与后期的实施控制，满足高端成功人士推崇的极致奢雅生活模式。

Zhongqiming, design director of Yijing Interior Design Agency in Chengdu. Huangying, chief designer of Yijing Interior Design Agency in Chengdu. Chengdu Yijing for indoor space design, interior design firm was main forced by the famous designer on a design. Since its establishment, it participated in a number of large indoor and outdoor design projects at home and abroad such as office, department stores, restaurant, hotel, finance, entertainment, clubs, exhibition hall, villas, high-end real estate etc, and won the prize in a number of domestic and international design competition, works published serially and distributed nationwide were by several domestic core media, website, magazines and journals widely acclaimed by customers. And now is also committed to high-end design and in prophase development and the implementation of the control, to meet the high-end successful people' s extreme luxury elegant life style.

Acknowledgements

We would like to thank all the designers and companies who made significant contributions to the compilation of this book. Without them, this project would not have been possible. We would also like to thank many others whose names did not appear on the credits, but made specific input and support for the project from beginning to end.

Future Editions

If you would like to contribute to the next edition of Artpower, please email us your details to: artpower@artpower.com.cn